Artificial Beings

Artificial Beings

The Conscience of a Conscious Machine

Jacques Pitrat

ISTE

WILEY

First published 2009 in Great Britain and the United States by ISTE Ltd and John Wiley & Sons, Inc.

ISTE Ltd
27-37 St George's Road
London SW19 4EU
UK

www.iste.co.uk

John Wiley & Sons, Inc.
111 River Street
Hoboken, NJ 07030
USA

www.wiley.com

Library of Congress Cataloging-in-Publication Data

Pitrat, J. (Jacques)
Artificial beings : the conscience of a conscious machine / Jacques Pitrat.
p. cm.
Includes bibliographical references and index.
ISBN 978-1-84821-101-8
1. Artificial intelligence--Moral and ethical aspects. 2. Conscience. 3. Consciousness. I. Title.
Q335.P558 2009
006.3--dc22

2009005292

British Library Cataloguing-in-Publication Data
A CIP record for this book is available from the British Library
ISBN 978-1-84821-101-8

Printed and bound in Great Britain by CPI Antony Rowe, Chippenham and Eastbourne

FSC
Mixed Sources
Product group from well-managed forests and other controlled sources
Cert no. SGS-COC-2953
www.fsc.org
© 1996 Forest Stewardship Council

Contents

Acknowledgements

First and foremost, I wish to thank Basile Starynkevitch and Carla who carefully read the first versions of this book, and suggested many improvements.

I also want to extend my gratitude to Robin who has been the system administrator of my computer since I retired from the CNRS. He has managed it so well that everything has run smoothly during all these years.

Work on this system began while I was a CNRS researcher until my retirement in 2000. I worked at the LIP6, a laboratory of Pierre and Marie Curie University (Paris VI), and I wish to thank all the members of the laboratory who assisted me in various ways during the first years of the development of my system. I am forever grateful to the members of the Métaconnaissances research team who provided intelligent comments and useful criticisms to my work.

I am grateful to the editors and readers of ISTE Ltd whose advice and suggestions were most helpful.

Note on the Terminology

Artificial beings (in contrast to human beings) are the subject of this book. As up to now most of their activities have been performed by humans, an extensive vocabulary can be used to describe actions when the performer is human, but few words are available for when it is an artificial being. Therefore, I will speak of artificial beings which think, choose, prefer, are conscious, want, etc. However, when I say that an artificial being "thinks" that only means that it processes some data so that it generates new data, which will help it to determine what it will do. For instance, it gives up its present task because it is too difficult, it creates a method for solving problems, it chooses the next rule that it will execute, it tries to understand the reason for a previous mistake, etc.

There is no anthropomorphism intended in this use of the vocabulary. It is simply due to the fact that it is impossible to create a new word for each activity of an artificial being when it is in the same situation as a human being, and when its results are often as good as ours. This does not mean that I believe that an artificial being operates as a human does. On the contrary, I will try to show that they can process information in a completely different way, particularly when they observe their behavior, which gives them a large advantage over us.

Let us consider an example: when I speak of the artificial being which I built, I say that it is autonomous. When we say that a human being is autonomous, that means that he can make his decisions without asking for help. In this meaning, my creature is certainly autonomous, it works without any interference for more than one month. However, this characteristic is not enough; a traffic light can work alone for years, it does not need an exterior agent to go from green to orange or from orange to red; an autonomous agent must not be completely predictable. However, this point raises a problem, because artificial beings are usually completely determined, they are following a program and, theoretically, we can predict all of their future decisions when we know their program and their data. Naturally, this comparison is unfair because we do not know how our brain works, even if it is as determined as a computer program. We cannot reject the possibility of an autonomous artificial being only because we do not know how human beings work. Moreover, theoretically predictable does not mean practically predictable. It is impossible to predict the behavior of a complex artificial being, it depends on its history and on a huge number of small operations, we will never have enough time to execute them. At different instants, it can take different decisions although it is in exactly the same situation because they also depend on the events that happened between these instants. The behavior of an artificial being may also be more difficult to predict than the behavior of a human one. For these reasons, I will speak of autonomous artificial beings, although it does not choose its act in the same way as us.

Finally, in Artificial Intelligence (AI) it is important to indicate unambiguously whether we are considering a human or an artificial being, particularly when we use pronouns. Indeed, both kinds of beings have similar activities and the context does not always indicate the nature of the agent. Thus, I will always use "it" to refer to an artificial being, which will often be my system CAIA. When I consider an activity in common with human and artificial beings, the agent is often represented by "one" as in "Let us assume that one has decided to interpret a program"; all we know of this being is that is able to make a decision, and it is not specified whether this agent is human or artificial. The passive form is also convenient to mention a being without specifying its nature of an actor as in "when the program has been written". In some situations, to insist on the fact that I am simultaneously speaking of human and artificial beings, I will write his/its, he/it, or who/which. I will never use his/her/its, the situation is complicated enough with two kinds of actors: when I introduce a human being, he may be male or female in all this book, even when I use "man" or "he". "We" will always represent human beings, the reader–author pair as in "We will later see that ..." or all human beings as in "We do not like the idea that artificial beings could become more intelligent than human beings".

Chapter 1

Presenting the Actors

At the beginning of a book, it is natural to present its actors. Of course, the first actor is the book itself. Then I will examine the qualities and drawbacks of two important families of actors: human beings and artificial beings. I do not forget the computer, an essential support for an artificial being. Next, I will present my reasons for developing my present research, and introduce the subject of this research: my colleague CAIA, which is an artificial scientist. Studying its behavior and its mechanisms will supply most of the examples of this book. Finally, I look at the domains where CAIA is carrying out research.

1.1. The book

The goal of this book is to examine whether an artificial being can have some capacities similar to those that consciousness or conscience give to human beings. Over many centuries, a huge amount of work has been done on human consciousness and conscience, but sometimes I feel that I still do not understand them. In relation with my work, to understand a hypothesis on the working of our brain, I need to see how the hypothesis can be implemented in a computer program. Naturally, scientists who lived before the computer era could not express themselves in such a precise way because they did not know of computers and how we can program them. This is a difficult and unnatural way of thinking, and even now, we have to realize several Artificial Intelligence systems before we can easily feed computer programs with attractive ideas. When I read these early books, I can see a dozen ways to implement each of their ideas and so I feel confused: among all these possibilities, which one was seen by the author? For instance, I have never understood how the qualia could be exactly represented. The qualia are at the core of many theories of consciousness, but philosophers agree neither on their meaning, nor on their properties. We can roughly say that their goal is to represent in our brain sensations like redness, but when the authors want to be more precise, each one has his own definition. How could we use this concept in a computer program?

Over several years, some outstanding books and papers, such as those of Marvin Minsky, Daniel Dennett, Gérard Sabah or François Anceau, have been written by scientists who know exactly how a computer works, and the reader can understand how their ideas might be implemented. However, these ideas have not always been effectively implemented. An AI scientist knows that it is impossible to foresee in a program all the elements that would be necessary to give excellent results: either some mechanisms are forgotten, or their description is not sufficiently accurate. We have to experiment with the system, and modify it to work better. Thus, a description without an implementation is an indispensable starting point, but it is not sufficient, although it contains many interesting and promising ideas. Moreover, the goal of most of the research has been to improve our understanding of consciousness and of conscience for human beings. Usually, they were not concerned to see if these faculties could be expressed by artificial beings, which have different ways to process information. However, there are some exceptions; in particular, John McCarthy has written a theoretical paper on the possibility of making robots conscious of their mental states. My book complements his work and I do not consider the theoretical point of view, but rather its realization by a practical computer system. His paper must therefore be read by all those who are interested in this approach. Several papers have been published in the last years on the realization of various aspects of consciousness. They contribute new and interesting ideas, but this is not enough: intelligence is a whole, and we cannot examine only how some of its aspects can be implemented, we have also to study what kind of consciousness has been given to artificial beings which have to perform difficult tasks.

Examining existing artificial beings

Another approach is to examine an existing artificial being, and consider whether some of its possibilities could be related to consciousness and conscience as they exist for human beings. Since, in a program, everything has to be defined, we can find which mechanisms generate a behavior similar to our own behavior. Besides, when an artificial being uses these mechanisms, it may have possibilities out of reach for us, whose performance depends on the structure of our brain and the characteristics of its basic element, the neuron. Then, we are no longer interested in the understanding of human cognition, but in the study of an artificial cognition. Its main goal is the realization and the understanding of artificial beings, for them to become as efficient as possible. In that situation, we have a huge advantage over the psychologists: we can examine all the programs that make up an artificial being, thus we can exactly know the reasons for its actions and for its limits. However, there is a practical problem because we may have to examine programs with several hundreds of thousands instructions: we may misinterpret these programs if we are not their author. Papers in journals may give a sketchy description, but thousands of pages would be necessary for an accurate description. Thus, a paper, and even a book, can only give a general idea of the methods used by an artificial being; the details cannot be included, although they are often essential to understanding its performance.

So, the only person who can accurately describe the properties of a system is its author. Several artificial beings have already used interesting mechanisms which give them some capacities related to consciousness or to conscience, one of the most impressive is Lenat's EURISKO. Unfortunately, I cannot describe them as much as I would because I cannot examine their programs in detail, I cannot know how some points have been dealt with, I cannot make experiments in order to evaluate their possibilities accurately. For this reason, I will mainly take my examples from the CAIA (Chercheur Artificiel en Intelligence Artificielle: an Artificial Artificial Intelligence Scientist) system, which I have experimented with for more than 20 years. CAIA has several attributes in common with human scientists, who discover new methods for solving problems, and perform many experiments to do so. However, CAIA does not have to deal with other important roles of current human scientists: CAIA does not publish scientific papers yet, CAIA does not search for scientific grants or funds, CAIA does not interact with the scientific community or policy makers, CAIA does not manage or advise junior human researchers.

CAIA is a step toward the realization of an artificial AI scientist. For the present, its main research domain is solving problems defined by a set of constraints, which must be satisfied by the solution. This family of problems includes many applications, and we often have to solve such problems. For instance, when we are choosing a meal for guests, we have to take into account a lot of constraints: their likes, the contents of the fridge, our cooking tools, our budget, the food supply at the nearest supermarket, etc. Crosswords and Sudoku problems are also in this category. My goal was not to develop an artificial consciousness or conscience; it was to realize a system able to learn to solve problems without the need to imitate the human behavior. Once CAIA was successful, I looked for and analyzed the mechanisms related to consciousness and conscience. When they were present, they were necessary for the success of the artificial scientist.

Plan of the book

This book comprises ten chapters. In Chapter 2, we describe some possible meanings for consciousness and conscience. Then in Chapter 3 we show that the concept of an individual is different for a human being and an artificial being. Chapters 4 to 6 examine what the consciousness of an artificial being could be: we describe several ways it can observe itself, why it is useful to observe its own behavior and how that can be implemented. Chapters 7 and 8 show that an artificial being can and must have a conscience. In Chapter 9, we examine some problems related to the conscience such as the importance of emotions, the difficulty in modifying its own conscience, and also the consequences of the existence of artificial beings on the human conscience. Chapter 10 discusses what the future has in store for CAIA. Appendix 1 describes the methods used for solving the problems given to CAIA while Appendix 2 gives more details of the implementation of CAIA's consciousness.

1.2. Human and artificial beings

Two of the most amazing features of human behavior are consciousness and conscience, which philosophers and psychologists have tried to explain for many centuries. How is it possible to observe a part of the processes that occur in our brain? How can we acquire and use the knowledge of what is right and wrong? These problems are difficult because we do not have access to the processes which are executed in our brain: if we could observe our mechanisms, it would be easy to answer these questions. Besides, no one can observe what happens in the brain of another person. Finally, although some experiments might help to understand what happens in the brain of a person if some of its parts were destroyed, they are naturally forbidden for ethical reasons. Therefore, studies on the working of the brain are very limited, and this helps explain why there are so many theories, which are often contradictory. However, everybody agrees that these processes are essential for satisfactory behavior. Socrates' motto was "Know Thyself", which can be realized because we are conscious. Furthermore, societies can survive only because the conscience of their members incites them to perform useful acts, and forbids them from committing actions dangerous for the future of the society.

Realizing artificial beings

We, AI scientists, realize computer systems, which are artificial beings. For the same reasons that consciousness and conscience are useful for human beings, we are led to give them the possibility to behave as if they were conscious and as if they had a conscience. The goal of some scientists is to model how the human brain works so that they will have a better understanding of human behavior. Indeed, if a mechanism generates a behavior similar to ours, it is possible that we are also using it. On the other hand, if the two behaviors are completely different, our brain certainly does not use the same methods as those implemented in the model. This approach is very interesting, but the goal of other scientists, and I am one of them, is not to understand how we work, but to realize systems that are as efficient as possible: we no longer claim to model the human brain. Naturally, a useful heuristic is to start from a hypothesis on how our brain works, but we are not obliged to follow it if we find better ways for performing the tasks given to our system. Naturally, once a system has been completed, it is interesting to analyze its methods, and perform some experiments on human beings to find whether or not we are using these same methods. It is also important to analyze its results for building an artificial cognition, which includes the description of new ways for implementing consciousness and conscience.

Constraints coming from the use of neurons

We must not blindly imitate human behavior because computers are tools which work differently from our brain. An obvious advantage is their speed, the number of operations executed in one second may be approximately the same as the number of operations that a human will execute in its whole life; besides, the computer will not make the thousands of mistakes made by a human. This is true when we are working in a serial mode, that is we cannot make several operations simultaneously, as is the case in arithmetic where we cannot make two divisions at the same time. However, the brain can also work in parallel, where it processes much data at the same time, for instance when we are perceiving a picture; for such application, the advantage of computers is not so large. Yet, the slowness of the neurons is not the only drawback coming from their use: we can neither observe their state, nor create new neurons. Another drawback is that it is not easy to store specific information such that it could be correctly used immediately, we have to use it several times before it is operational. I know that cars drive on the left in England, but I must be careful when I am in Great Britain because this information is not available each time that I cross a street. After a few days, this becomes automatic; unfortunately, when I return to the continent, it will be almost as long to acquire new reflexes. For a computer program, it is sufficient to swap "right" and "left" in a program, and everything is adapted in the new environment immediately; it is also easy to return to the previous behavior by doing the inverse swap.

Constraints coming from the structure of our brain

However, some characteristics of human intelligence do not come from using neurons, but from the structure of our brain. For instance, our working memory can hold around seven elements, which is a serious restriction for many tasks. We are not aware of the consequences of the small size of this memory because every human has this handicap; we avoid putting other people in situations where a larger working memory would be necessary. For instance, we speak in such a way that our interlocutor can process our sentences. We could possibly generate sentences such as:

The mouse that the cat that the dog that the woman that the soldier loves pets hates devours.

However, our interlocutor cannot understand it, while he can easily understand the following equivalent sentence:

The soldier loves the woman who pets the dog which hates the cat which devours the mouse.

We cannot stack in our working memory enough items so that we can wait for the many verbs that are at the end of the first sentence. Our brain evolved so that our ancestors, hunters and gatherers, could survive, breed and raise their children. The natural selection has created a wonderful tool for all of these activities, where a large working memory is not essential. Therefore, evolution has not developed this kind of memory, and it is the same for many other activities such as proving mathematical theorems, or managing large organizations. Our intellectual powers were useful for our ancestors' activities, and it happens that some of these powers are also very useful for other activities. However, there is no reason why they would be optimal for these new applications: the structure of our brain has not been optimized for them. Consciousness and conscience are among those useful aptitudes which evolution has developed for our ancestors, but it is likely that it would be possible to improve them so that we would be more efficient for most of our new activities. In fact, with the restrictions due to the neuron and those due to the structure of our brain, we are handicapped in comparison with artificial beings which are faster and which can simulate a huge variety of mechanisms.

Therefore, when we want to realize an efficient artificial being, we do not restrict its performance by imitating our way of thinking, thus enforcing constraints that are not essential. On the contrary, we have to take advantage of their extraordinary possibilities so that we obtain the best results. The goal of this book is to show that several features of consciousness and of conscience may be improved in that way. However, some of the human capacities have not been given to CAIA because they did not seem useful for the tasks which it is doing at present. Naturally, for future developments, it will certainly be necessary to give CAIA some of these missing capacities.

1.3. The computer

The computer is the extraordinary tool that enables an AI scientist to implement and experiment with his ideas. The methods which will be given to an artificial being are first defined as programs and data, then they are given to a computer which will execute these programs and give the desired results. We do not need exceptional computers for performing most AI research. I use an ordinary PC, such as those which are sold in their millions each year. It is not ultra-fast (1.8 GHz), and it only has 512 MB of memory.

However, a computer is almost useless on its own, we need a program to manage its working, which is the operating system. These systems did not always exist: the first computers were used without an operating system, but the present computers are so complex that it is almost impossible to use them without an operating system. It manages the computer resources, it decides which programs will be executed (a program being executed is called a process), it allocates them memory chunks, it deals with the inputs made through the mouse and the keyboard, the outputs such as those on the printer or the monitor, the network connections with other computers, and so on. It also checks that several parameters necessary for a satisfactory use of the computer are correct, for instance it will start a fan if it finds that the temperature inside the processing unit is too high. I am using Linux as an operating system.

The behavior of an artificial being is defined by a set of programs and data. To start it, one gives it to the operating system of a computer, which will create a process corresponding to an execution of this program. A process contains data and code expressed in a language understandable by the computer. In this book, I will distinguish the program, which is a sequence of instructions given to a computer with its data, from the corresponding process generated by the operating system controlling the execution of this program.

Thus, the operating system manages many processes, including the process corresponding to the execution of our AI program, and the several processes that our original process may have launched when needed. Even when we have given no task to our computer, 30 or so processes may be managed by the operating system, for its own needs or possibly for other users of the same computer. As many processes are simultaneously controlled by the operating system, it must avoid any interference between them: if our process were to write in an area of the memory used by a process belonging to another user, this last process would give the wrong results. So, the operating system allocates an area of the memory to each process; if an instruction of a process requires use of a memory which is not in this area, the computer will automatically realize that there is a mistake; it will reject the execution of this instruction, and a message is sent to the guilty process. Everything happens as if our process was the only user of a computer with a memory restricted to the area allocated by the operating system.

1.4. The author

My thinking about consciousness and conscience comes from CAIA, which I have been developing for more than 20 years. I wanted to realize a performing general system and, at the start, my goal was not at all to study the consciousness and conscience of artificial beings. It was only when the system began to have satisfactory results that I wondered whether some of its behavior had some aspects similar to human consciousness and conscience. So, all the characteristics which are relevant to these domains were not introduced to imitate human behavior, but because they were necessary to improve efficiency.

I started developing CAIA because I was struck by the slowness of the development of AI. Although I believe that it is certainly possible that artificial beings could be more intelligent than human beings, I have come to doubt whether human intelligence alone could ever realize such a difficult task, without exterior help.

Bootstrapping AI

Thus, we need help, and who could be intelligent enough to help us? The only possible candidate is AI itself; we have to build AI systems which can help us to develop AI. This means that we must bootstrap AI. The idea of a bootstrap is paradoxical: how can a system be used to implement itself? Actually, there is a succession of artificial beings, and system number N helps us to implement system number N+1. If system N+1 is better than system N, then it will help us to realize system N+2 which will be better still than system N+1. The bootstrap is based on the collaboration of a human and a system in order to produce a better version of this system. In that way, we realize a sequence of artificial beings; as each one is better than its predecessor, either the steps are higher or the human help can be decreased. For CAIA, the value of N is around 1,500 at the present time. So, it is completely different from its initial state in 1985, with much more than 1,000 intermediary versions, each one being used to generate the following one. I hope that, eventually, artificial beings will create better artificial beings without any human help; in that way, AI could progress by itself but, at present, we are very far from that state. Bootstrapping is often used for solving difficult problems, for instance we could not have realized the present computers if previous computers did not exist. The very first computers (in the 1940s and in the 1950s) were very simple, and could be made by human engineers without the help of a computer. As computers grew more and more powerful, they have been used to design still more powerful computers, and the design of current computer systems (from the microprocessors to the operating system and application software) would be unthinkable without a wide use of computers.

The key step for bootstrapping AI is the realization of an artificial AI scientist. Indeed, once such a scientist would be completed, it could develop AI without any human help. Naturally, it could also develop all of the other domains: mathematics, computer science, physics, management, etc., since AI is useful in all these domains. Besides, it will not only build methods for solving problems in these various domains, but it will also improve itself as an AI scientist; so, it will become a more and more competent AI scientist, and all the possible applications will benefit from these improvements. Thus, I started to develop the first steps of this venture. At the beginning, it is illusory to try to immediately create a scientist as able as a human one in every domain: it can only carry out a part of the necessary tasks, and I help it for the other ones.

In a bootstrap, a human has a crucial role, because a bootstrap is a sequence of steps from one version of the system to the following one. Someone has to decide when a new step will be made; to do that, he analyzes the results, and makes the modifications to the current version of the system in order to define its new improved version. I call this person the leader of the bootstrap, and I am the leader of CAIA's bootstrap.

I did not want to ape human methods, but I had to give my artificial scientist some possibilities related to those given to us by consciousness and conscience. Unlike what happens for human scientists, we can observe the inside of an artificial scientist, and see how these capacities have been implemented; due to this advantage, artificial cognition is easier to develop than human cognition. I take most of the examples from my experiments with my artificial scientist, which has already lived several lives, each one for more than one month, without any outside intervention; moreover, a bug has never stopped it. It would be possible to have longer lives but, when I examine its behavior and the events that happened in one month, I have many ideas for improving it. It is therefore better to make these modifications, and to launch it in a new life where its initial capabilities have been improved. I will mention other works on artificial beings, but I often prefer to speak of CAIA because Know exactly how its modules work.

1.5. CAIA, an artificial AI scientist

When we say that a person is conscious, it is difficult to explain this clearly because the restrictions of our consciousness forbid us to know what happens when we are conscious. This is why papers on consciousness are sometimes not very precise, which seems almost acceptable because we already have an idea on the subject. However, the situation is completely different when we are considering artificial beings, we can know everything of their knowledge and of the steps executed for achieving a task. This is the case for CAIA, my artificial scientist in AI. We work together to realize better AI systems, each one of us performing the tasks at which he/it is the best. While CAIA improves, there are more and more tasks that it can accomplish at least as well as myself. The goal of the completely autonomous AI scientist is very far away but, even now, CAIA is very helpful for many activities.

CAIA, a useful collaborator

CAIA is permanently evolving, on the one hand by itself, and on the other hand through the modifications that I make. I try to give it some of my activities so that I can spend more time on those that it is not yet able to perform. At the beginning of a bootstrap, we simplify the tasks given to the system so that it can achieve them correctly, although it does not yet have all the capacities expected for the final system. One simple way to help the system is to limit the nature of the problems that it will have to solve. This is why I gave it at the beginning only constraint satisfaction problems; now I am adding the capacity to solve arithmetic problems, and other domains will follow. Another way to help it is to ask it to solve only a part of a problem. For instance, when it has to solve a constraint satisfaction problem, I do not give it the formulation in a natural language such as English or French, it is instead given in a formalized language, similar to the mathematical language in many points. This removes several difficulties: for instance, it is no longer necessary to clear up ambiguities that are always present in a natural language text. In the distant future, when it will be able to solve problems expressed in natural language, another step of the bootstrap will have been made.

As I have already said, my artificial scientist has already had several lives, which lasted more than 3 million seconds each, that is more than one month, night and day. This is huge for a system that runs on a computer which executes more that a billion instructions each second. During each of its lives, it is completely autonomous, I do not step in, I never even examine what is happening. It creates several methods for solving problems, it uses them to solve problems, it creates new problems, it performs some experiments to see whether it could be possible to find better solutions with other methods, it analyzes these experiments (but it does not interpret these analyses, this is still my role), it tries to understand why one of its methods did not lead to a better solution for a problem, etc.. Of course, during this time several bugs occur, a complex program without bugs is a purely theoretical ideal. In such a situation, it finds which instructions are wrong (but it does not correct them, this is also still my role), and if the bug had put it into a loop, it will break this loop. I generally stop CAIA after about one month because I then have more than enough improvements to make from observation of its behavior during this period. It is more interesting to modify it, and observe what will happen in a new life.

Thus, I am progressively increasing the abilities of the artificial scientist, which already helps me because it now performs some of the tasks that I had to make at the start. Some of its activities are related to consciousness and to conscience. First, it observes what it has done so that it can understand the reasons for its successes and its failures, it examines its own knowledge so that it can foresee some possible uses, it observes what it is doing so that it immediately recognizes that it is going in a wrong direction. To do that, it has some of the possibilities given by the consciousness. However, as it is autonomous, it also has the capacity to judge what it is doing, which is what conscience enables us to do. It can see whether some results are good or poor so that it can try to repeat or to avoid them later. It can decide whether a new problem, which it has just created, is interesting so that it can decide whether to keep it. It evaluates the interest of the waiting tasks so that it can manage its life as a scientist, beginning with those which are the most important and the easiest.

Should we imitate human behavior?

Although my goal is to realize an artificial scientist, I am not bound to imitate a human scientist. We have already seen that our cognition has some restrictions, which artificial beings do not have; it would be a pity not to use their capacities completely. Naturally, observing a human subject, mainly myself for CAIA, is very useful for finding initial ideas, but AI scientists do not reject methods that we could not execute, for instance because we would need too much time to do them. My initial goal was not at all to create an artificial being which would be conscious, and which would have a conscience. It only happened that it has some of these possibilities because they were necessary for its performance: it needs to be conscious to learn and to adapt itself to new situations, and it needs a conscience to be autonomous.

Most examples will come from these experiments, and I am often slightly envious of the possibilities of the artificial beings, which we will never have. We will see some of these in the study of the consciousness and of the conscience, but they also arise in other situations. For instance, an artificial being can not only replace a module insufficiently successful by a more efficient one, but it can add a new module specially tailored to a new application. Unfortunately, we cannot do that; for instance, our mathematical abilities are not as high as we could wish, evolution did not build us for that: this is only a by-product in the creation of modules enabling us to hunt or to seduce the opposite sex; we cannot add several billions of neurons specially organized to perform the tasks useful for a mathematician. On the contrary, an artificial being can possibly add to itself a large number of modules specialized for a new domain.

The main drawback of this approach is that we have no model that we could imitate, we could make many more things with a computer if we only knew that they would be useful. This permanently happens when I am developing CAIA, it can have activities that no human scientist could have but, as we have no model using them, it is difficult to define them. Moreover, we cannot find how human scientists work, because their most interesting processes are unconscious; the illumination of the discovery is well known, but what happened before this illumination? The existence of our unconscious mode of thinking is a severe handicap for the realization of artificial beings: we cannot use ideas coming from the observation of our behavior if we do not know how we work. So, we often have to wait for an unexpected success obtained by an artificial being, and then to understand what happened. It is when one analyzes the reasons for this success that new ideas can be found, which can later be included in its knowledge for solving problems. For this reason, it is essential to experiment; we cannot find many ideas on a blank sheet of paper. Luckily, an artificial scientist can make and analyze many more experiments than a human scientist, thus it is much more likely that it will have fortunate surprises. Due to this capacity to perform so many experiments, the collaboration of artificial scientists with human scientists has an enormous potential for progress, and this will help to bootstrap AI tremendously. For the present time, the work is shared in the following way: CAIA makes experiments, it analyses them, and I draw inferences from these analyzes for improving CAIA.

Choosing a research domain for the artificial scientist

In the first step, CAIA could only solve constraint satisfaction problems (CSP); the name seems impressive, but we are currently solving such problems in our everyday life for instance, as we saw at the beginning of this chapter, preparing a menu for our guests. Many puzzles given in the newspapers are CSP such as Kakuro, finding magic squares, knight's tour on a chessboard, eight queens puzzle, etc.; many industrial or scientific problems can also be formalized in that way. This is the case for the generation of new problems, which I have defined as a CSP problem; in this way, CAIA generated thousands of new problems. This is indispensable because a scientist has to give his/its system many problems so that he/it can experiment this system thoroughly. However, finding new problems may be difficult; one has to find them in various publications, and then type their data. It is easier if the artificial scientist does this task alone, that helps its human colleague. Moreover, many problems are often too easy, it is necessary to have very difficult problems so that they try out the possibilities of the solver generated by the artificial scientist. Thus, the choice of a domain which enables it to create new problems is an important help to the human scientist, which delegates in this way a part of his work to his artificial colleague. In a second step, I am giving CAIA a new domain: solving arithmetic problems.

1.6. The research domains of CAIA

In this section, we temporarily move away from consciousness and conscience because it is impossible to describe the working of an artificial being abstractly. It is not sufficient to describe toy problems, which finally prove nothing, any method can succeed to solve easy problems; only examples of difficult problems can clearly show the interest and the difficulties of this approach. Moreover, usually consciousness is not essential for solving easy problems.

Choosing a research domain for CAIA was crucial because this domain has to include many difficult problems, otherwise it will not develop its capacities for research, and it will not show its qualities. Its first domain was solving Constraint Satisfaction Problems (CSP); they are defined by a set of constraints, which must be satisfied by the solutions. Many problems given in journals in their games column can be described in this way, but many other industrial or mathematical problems can also be stated as constraint satisfaction problems. Some of the problems solved by CAIA are really difficult, and the first human who solved some of them published their solutions in a scientific journal. Many people prefer to solve easier problems, such as Kakuro or Sudoku, and CAIA is often able to generate and solve problems of the same family as those given in the journals, but too difficult for human solvers. We have already seen that this domain is so extensive that it includes problems useful for solving some needs of CAIA, such as the generation of new problems. In this way, the ability to solve CSP problems is useful for developing the system, several tasks are solved by the solver of CAIA instead of being solved by myself. This is an example of a virtuous circle, which frequently occurs in a bootstrap: if CAIA improves its performances in problem solving, it can create better problems which will be useful for improving its performance, etc.

Crosswords, an example of constraint satisfaction problem

Among the many CSP problems, let us consider crosswords where there are three kinds of constraints: the word must correspond to its definition, the number of its characters must be the same as the number of squares at the place where it will be put, and finally, when two words have a common square, the letter for this square must be the same for both words. Usually, many words are possible for a definition, even if the number of letters is taken into account. For instance, for "French town" with five letters, we have Paris, Lille, Brest, Rouen, etc. However, it may happen that all of the candidates have the same letter at the same place. This is not true for the preceding example, but if the definition is "French capital", with five letters, there are few towns: Paris, Blois and Tours. We can check that, for these three towns, the last letter is "S"; thus, this character can be written on the corresponding square because it is certainly right, although we do not know which town is the correct one. Unfortunately, it is not always possible to be sure of a choice, one has to take one of the candidates, and write it on the grid. After that, several possibilities will be removed, other words will become certain and, if one is lucky, one will complete the grid. Unfortunately, the choice is often wrong, and later on one finds a contradiction: no word can be put at some place. Thus, with an eraser, one deletes one of the preceding choices, and also all the words that have been placed following this choice; then one chooses another possible word for this place, and one writes it.

An analogous method can be used for solving all of the constraint satisfaction problems: one makes a choice, one draws inferences from this choice, and one removes the choice when there is a contradiction. The step where one removes a preceding choice and all of its consequences is called a "backtrack" frequently occurs when one is solving a difficult CSP problem: one often backtracks, for instance the use of the eraser corresponds to a backtrack. A skilful problem solver judiciously determines the item that it will choose, and he/it draws many inferences from each choice. With these possible choices, one generates a tree; a good problem solver generates a much smaller tree than a poor one, so his/its solution is much easier to explain.

We can apply several rules in order to solve a crossword puzzle. We have already seen one of them: if the same letter will be put on a particular square for all of the possible choices, then one can put this letter on this square; this enabled us to put "S" for the fifth and last letter of the word corresponding to "capital of France". Another rule indicates that, if no word can be put at a particular place, there is a contradiction: one must backtrack. A third rule: remove a candidate if its letter for square S is different from the letter already written on square S; if letter "R" is written on a square which corresponds to the second letter of the word in the perpendicular direction, all the candidates that do not have "R" as second letter are removed. CAIA does not solve complete crosswords puzzles because it is not yet able to understand natural language. However, it works in the same way: it generates candidates, it applies the inference rules, it chooses an item among the candidates, it backtracks. In addition, it has many more rules, and its rules are much more general than the preceding rules since it can solve many kinds of problems.

A general method for solving CSP problems

We will summarize a general method for solving CSP problems: one applies rules which clarify the situation; removing impossibilities such as a candidate certainly not satisfying a constraint, or finding a part of the solution such as the value of one of the unknown elements, or generating a new constraint easier to use than the original constraints. If no rule can be applied, we choose a value for an unknown item among its possible values, and we resume the application of rules, using this new information. If there is a contradiction, one backtracks.

A successful problem solver knows many rules, so he/it has to choose among them those which can advance toward the solution, and does not waste time with useless rules; he/it has a method for making these decisions. A poor problem solver systematically tries all the possible candidates until it finds a contradiction, in a blind combinatorial search. This kind of method is also often used by computers because they are so fast that it works even when millions of choices have to be considered. For this reason, computers have a bad reputation, but this is the fault of the scientist who has not built an efficient method for solving problems. When an artificial being has an efficient method for solving problems, if will find solutions as elegant as those found by clever human problem solvers. For this reason, I call it the clever method. Such a method associates with each rule three elements: a trigger, a set of conditions and a priority. The trigger indicates that it may be interesting to consider this rule: if we are hungry (which is the trigger), we consider the possibility to go to a restaurant. The conditions decide if a rule which has been triggered will be executed: for going to a restaurant, we must have money. Finally, the priority indicates if this rule will be immediately executed, or after more interesting rule, or only if there is nothing else to do: considering to go to a restaurant has a medium priority, this is not the most usual way for eating. A human scientist, as well as an artificial scientist such as CAIA, discovers clever methods, which give clever solutions when these methods are used by its solver. Thus, this solver often develops very small trees, it often finds solutions as elegant as those found by very good human problem solvers. One of the main goals of CAIA is to build a general and efficient method for solving problems.

We will see that CAIA may be called "conscious" according to several meanings of this word. In particular, it can observe its behavior, which enables it to monitor its solver. In that way, it can improve the solving method it has generated, evaluate its performance, guide the solver, help it to make good choices when it has to find a value for an unknown item, explain the solutions, and even explain how it has found a solution. It is also able to manage its activity autonomously as a scientist does: its "conscience" helps it to find whether its actions are right or wrong, whether the new results that it finds are interesting or unimportant.

For the reader who wants to go deeper into CSP problems, in Appendix 1 I describe other problems, and give more information on the rules that can be used for solving them. These examples require only knowledge of the four basic operations: addition, subtraction, multiplication, and division. Human beings usually know how to solve these problems, but we are not accustomed to think about the methods that we are using. This is why I give there a broad outline of CAIA's methods, which can be used by human beings as well as by artificial beings.

Arithmetic, a new domain for CAIA

CAIA has now a second research domain: arithmetic. An example of a problem solved by CAIA is the Pharaoh's problem, where one has to find four positive integers X, Y, Z, and N such that

N is a prime number

$X^2+Y^2=Z^2$

$X*Y*Z=348,960,150*N$

This difficult problem only has one solution if we take into account the symmetry between X and Y; the main lines of the resolution found by CAIA are given in Appendix 1. Contrary to CSP problems, the variables have an infinity of possible values. Thus, it is no longer possible to use a combinatorial method where one considers all the possible combinations for the values of these variables.

I chose the arithmetic domain because the manipulation of numbers is important for a scientist using computers. Moreover, this leads it to tackle theorem proving, also a crucial activity for a scientist who rigorously proves the properties that it will use for improving the performances of its system. Another interesting domain would have been computer science, which is central for a scientist which is itself a computer system. I hesitated a lot before preferring arithmetic, I thought that giving it some abilities in theorem proving would be useful soon for its own needs, but computer science is probably the third domain where CAIA will have to work.

1.7. Further reading

On the qualia, Wikipedia presents several points of view according to various scientists. In *The Society of Mind*, Marvin Minsky examines in several sections how consciousness is implemented in the human brain, particularly in Chapter 6 "Insight and Introspection" and in Chapter 15 "Consciousness and Memory", but the importance of consciousness appears everywhere in this book, see also 19.1 "The Roots of Intention" and 28.8 "Overlapping Minds". In *The Emotion Machine*, he adds many deep remarks on the subject, especially in Chapter 4 "Consciousness", and Chapter 9 "The Self". Daniel Dennett's book, *Consciousness Explained*, must be read by anyone who wants to understand how the consciousness of human beings can be described with computer models. In *Vers une étude objective de la conscience*, François Anceau presents a computer model of consciousness, another model is studied by Gérard Sabah with CARAMEL which is described in several papers such as *Consciousness: A Requirement for Understanding Natural Language.*

John McCarthy presents his striking theoretical study of how robots could become conscious in *Making Robots Conscious of their Mental States.*

Several books are focused on the consciousness of artificial beings. Two books are a collection of papers on this subject: *Machine consciousness* edited by Owen Holland, and *Artificial consciousness* edited by Antonio Chella and Riccardo Manzotti. Moreover, 25 papers were presented at the 2007 AAAI Fall Symposium on AI and Consciousness. Some of these papers are theoretical, others suggest models or projects for realizing conscious machines, and several realizations are also described. In particular, in the first book, Luc Steels starts from his experiments where a population of robots invent a language; he studies what kind of information processing structures can create an inner voice, that is the silent voice that we are often hearing when we are not speaking. In *The cognitive approach to conscious machines*, Pentti Haikonen describes how consciousness could appear in neural networks, which are inspired by the structure of the human brain; in *Robot brains: Circuits and systems for conscious machines*, he gives plausible proposals for building conscious robots.

A recent description of CAIA is in Pitrat, J. *A Step Toward an Artificial Artificial Intelligence Scientist.*

The following papers describe two impressive programs solving crossword puzzles:

– *A Probabilistic Approach to Solving Crossword Puzzles* by M. Littman, G. Keim and N. Shazeer;

– *A Web-Based Agent Challenges Human Experts on Crosswords* by M. Ernandes, G. Angelini and M. Gori.

Many arithmetic problems are given at:

http://www.diophante.fr/pages/arithm2.htm#hautdepage.

The Pharaoh's problem which I mentioned is problem A467 on this site.

Chapter 2

Consciousness and Conscience

Why is it useful to study consciousness and conscience simultaneously? This is not so surprising: although these concepts are different, they have a common root, the Latin word "conscientia". Latin is not the only language where both concepts are represented by only one word, this also happens in French. This link, which appears in some languages, has an explanation because both concepts are related in several ways. In particular, consciousness gives us some information on our actions, including why we are doing them; this is useful for the conscience, which will decide whether they are right or wrong. Thus, the same action is evaluated by the conscience according to the goals of its author. For instance, the surgeon who cuts the chest of a seriously ill patient is a helper, while the gangster, who stabs someone because he wants to steal his money, is a criminal. This is true even if both the patient and the victim died. When we want to appraise a behavior, we must know why it has been prompted, and this is in the head of its author. Conversely, the conscience plays an essential role for choosing among several possible actions considered in the conscious reasoning: some of the actions that satisfy a goal are accepted or rejected because they agree or conflict with our conscience. To rob a bank is an efficient way to get money, but we reject it even when our goal is to get a lot of money; our conscience does not accept it, even if it is not dangerous. Consciousness and conscience are at the heart of the mechanisms that build the behavior of a human being, and this is also true for an artificial being. Two key elements of an individual are the possibility of it observing its own mental processes, and of monitoring them according to his moral principles. Usually, the law considers that a human being is not responsible for his offences if he was not conscious when he committed them.

For their consciousness as well as for their conscience, human beings have possibilities incomparably better than those of animals, even including the cleverest ones such as apes, dolphins or elephants. However, for several years, we have been creating artificial beings; it is therefore interesting to examine their possibilities: for the first time, non-human beings may have possibilities similar to ours for consciousness and conscience. They may use mechanisms very different from those running in our brains, but they may obtain good results; perhaps, they can use excellent methods that we will never be able to use because of the specificity of our brain.

Firstly, we will consider several meanings of the words "consciousness" and "conscience", then we will see why it could be useful to give these capacities to an artificial being. Finally, we will wonder whether it is possible to create beings that would have capacities which will always be out of reach for us, due to the constraints coming from the realization of our brain. These new abilities could be very useful for improving the performance of artificial beings.

2.1. Several meanings of "consciousness"

Several meanings can be associated with the words "consciousness" and "conscience", and it would be useful for an artificial being to have some of the possibilities corresponding to these meanings. For some other meanings, either we do not yet know how to implement them, or we do not realize that they could improve their performance. Furthermore, new meanings may be found for these words, which are related to activities that are feasible for an artificial being, but that we could never have. It is difficult to analyze these words, each of us has a personal interpretation; these phenomena are not well understood because it is difficult to observe oneself, and still more difficult to observe other human beings. Of course, neuroscience has made spectacular breakthroughs, and some amazing psychological experiments have been performed; yet, much work has still to be done so that we may understand the mechanisms used by human beings when they are conscious of what they are thinking. We will see that the situation is clearer for artificial beings; we can observe them much more thoroughly than ourselves or another human being. Moreover, they can observe themselves in a more complete way than we can observe ourselves and they have many powerful ways to do so. We begin with several meanings associated with the word "consciousness".

Consciousness as perceiving the world

Firstly, there is phenomenological consciousness. It is linked to the sensations that our brain receives from the world and to the perceptions which result from our processing of these sensations. Then, we reason and decide some actions from this information. A more specialized word, "awareness", indicates that we can represent the exterior world in our brain. An artificial being also needs this kind of consciousness; it receives data by sensors, such as a keyboard, a mouse, a webcam, etc., and its programs will process these data. I will not deal further with this kind of consciousness, which all artificial beings have.

Consciousness as being awake

A second meaning is to be in a state contrary to deep sleep or coma, where we have lost consciousness. Usually, our brain is awake, we know that it is running, we perceive information from the outer world, and we act on it. As for the first meaning, this is also described by the specialized word, "awareness". This meaning is not of the highest importance for an artificial being, which cannot be knocked on the head, but we will see that it can be interrupted by the operating system or by the leader of the bootstrap; in this case, it is put in a state where no processing occurs, but it can be restarted exactly at the point where it was stopped. In this situation, it is evidently not conscious, but when it is running it is always conscious in that meaning, except when it happens to be in an infinite loop. When an artificial being is in such a loop, its processes execute billions of instructions, always the same, but none of them is in a module where it can observe what it is doing. That may last for ever, but we will see later that CAIA can break these loops. Naturally, as long as the system is in a loop, it is not conscious according to this meaning. In a following section, we will examine more thoroughly this possible meaning of consciousness for an artificial being.

Consciousness as being unique

A third meaning is the feeling of existing, of being unique, different from other beings. In that case, a specialized word is "self-awareness". This particularly appears in the mirror test, where someone has to recognize that the image in the mirror is an image of himself. Can he see that the image reproduces exactly all of his movements? If we mark a part of his body with a dye spot, will he try to remove it? After 18 months of age, human beings succeed in this test, but apes and elephants also pass it. Evidently, CAIA cannot succeed in these tests, since it has no physical image. A more serious limitation is that it has no idea that it exists and that there are other beings. It has some ideas of its abilities, but it has no model of its leader of the bootstrap. It does not even know that I exist, it only accepts my orders, and it tries to fulfill them. It does not know that I can modify it, but it executes my orders even when they will modify it. It would be possible to realize an artificial being which would have a model of itself and of other beings, but I did not implement that because it was not useful in the simplified world in which CAIA lives. In any case, it would be difficult to give it the impression that it is unique, when one can clone it instantly into millions of identical copies! How could it have the impression that it is different from the other beings when it could interact with many twins, which are more identical than human identical twins, since they share all the experiences that they had before being cloned?

Consciousness as observing its reasoning

Another meaning is access consciousness: the information in our mind is accessible for reasoning, and therefore enables us to control our behavior. This possibility of knowing some of our mental capabilities has many useful consequences. We can check that our inferences are correct. We can make a verbal report to other beings who will more readily accept our conclusions. We can consider which steps were taken for solving a problem, and know why we have chosen to make these steps; thus we can learn from our successes and failures. We can monitor our search for a solution: when we see that we are starting in a wrong direction, we immediately stop and we try to find a better way, taking into account the failure of the preceding attempts. Our consciousness has a huge importance for controlling our actions; however, this possibility of self-observation is limited for human beings, a large part of our mental processes are unconscious because we cannot gain access to them. We will show that artificial beings can have many ways to observe themselves, and that they are not as restricted as ourselves. They may have a more complete access to their processes while they are working; they can also know their present state almost perfectly, while human beings have almost no such capability. We will be mainly interested in this meaning of consciousness.

2.2. Extending the meaning of "conscience" for artificial beings

Let us now consider the meaning of the word "conscience". There are two points of view when we are speaking of conscience and of artificial beings. The first question is to know whether an artificial being may have a conscience, and whether it is useful; this will be the main point developed in this book. The second question is to determine whether we must modify our own conscience for controlling our relations with artificial beings; if so, we will have to add new laws in our conscience. For instance, can we destroy them, or use them as sexual slaves? We will briefly consider this point at the end of Chapter 9.

The human conscience

The meaning of the word "conscience" seems clear: it tells us whether what we are doing is right or wrong. It is important when we are choosing our goals and the actions that enable us to fulfill these goals. First, it favors the goals that agree with our conscience, and it penalizes those that disagree. Then, it also favors the actions that are allowed by our conscience. Finally, it acts even at the completion of a goal: if some of the consequences of our actions are forbidden, we feel an emotion, remorse, which urges us to obey our conscience in the future. The difficulty is to find how one can set up and respect what is right and what is wrong.

There are three levels of conscience: the first one is useful for an individual, the second one is useful for a group of people such as a country or the believers of a religion, and finally the third one is useful for mankind. Moreover, the conscience that benefits an individual may be split in two: one favors its survival and the other the survival of its genes. Evolution favors the survival of the genes, it secures a large offspring to a competent being; in the same way, the Catholic church considered that if, during a difficult delivery, one had to choose between saving the mother or saving the baby, one must choose to save the baby. However, the survival of the individual is indirectly favored by the fact that he must live to have a large offspring.

A human being belongs to many societies, and the conscience often defines goals and interdictions favorable for the survival of these societies. If a society has made a wrong choice for the laws which it imposes on its members, it will disappear. It is difficult to define these laws because a society can be suddenly immersed in a new environment and, frequently, it is not able to modify its laws fast enough. This may happen when a society begins to interfere with other societies, and this encounter brings about new problems, which cannot be solved with the old principles. The society may also be confronted with new diseases or with important climatic modifications, and it can survive only if it changes its laws drastically. When the highly sophisticated societies of the Aztecs and the Incas encountered the Spanish, they were suddenly fighting against people who had knowledge of firearms and horses, and who carried diseases which they could not cure. As a consequence, they were destroyed in a few years. In the same way, with the moral principles of several present societies, we have seen a tremendous progress in many technical domains. It is not evident that we will modify our moral principles fast enough to face up to the challenges arising from the increase of the world population and of its standard of living.

We do not include in our conscience rules that we are obliged to follow because they are so important that evolution made us in such a way that we cannot disobey them. For instance, it is unnecessary to enforce a rule stating that we must breathe; although this is crucial for our survival, we cannot stop breathing. There are also important actions which are not mandatory in our conscience, because evolution motivates us to do them adequately. For instance, it is important for a society that its members do not starve, and produce many children who will be its future members. All the same, the states and the religions do not require us to eat at least three times a day, or to have sexual relationships at least once a week: most people will spontaneously do it if they can. We are built in such a way that our body gives a signal when we infringe many rules that could be harmful to human societies or to ourselves: when we are not eating enough, hunger incites us to eat, when we are risking damage to a part of our body, we feel a pain and we immediately act so that it stops. We do not include these latter constraints in the conscience because we can neither create them nor change them, since they are innate. Thus, a large part of our conscience is related to principles and rules that are given because we do not always want to respect them although they are essential for the survival of a society. The law obliges us to respect a speed limit, because many people love to drive too fast, although it is dangerous for themselves and for other drivers.

The conscience of an artificial being

The separation between rules that we must follow and rules that we naturally follow only depends on our physical realization, not on the importance of these rules. There is no reason why a similar separation should exist for an artificial being, it is arbitrary. Many rules have the same goal, the survival of the individual or of a society of individuals, but they are implemented either by innate constraints or by laws and principles that we must respect. Our definition of conscience depends on our physical realization; our main goals are to assure our survival, the survival of our genes, and the survival of our societies; our conscience is only about the goals that we could try to ignore.

Furthermore, the human conscience is related to goals which we consider as important, such as avoiding death; we do not include in our conscience the goal of solving a problem. However, an artificial being is built in such a way that it is very serious for it to fail to solve a problem: our goals and their goals are not the same. This is a second reason why the conscience of an artificial being can seem very different from the human conscience. All the same, the conscience is related in both cases to the need to satisfy crucial goals.

Therefore, when I am considering an artificial being, I include in the description of its "conscience" all the elements that favor the goals and the actions defined as good by its leader of the bootstrap, that is, which enable it to satisfy his goals and which incite it not to commit actions which he considers as wrong. Some goals are the achievement of a task while other goals are needed so that the artificial being can correctly work; for instance, it must not damage itself. Artificial beings also have given laws, which they cannot modify, and learned laws. As this separation between these two kinds of laws is not always the same as for human beings, we have just seen that it is necessary to extend the definition of the conscience well beyond the definition usually used for us. I will consider that the conscience of an artificial being includes all the knowledge that enables it to fulfill its goals correctly: to achieve the goals given by its leader of the bootstrap and to avoid actions that would prevent it to work normally or would even lead to its destruction. In this way, I include in the conscience of an artificial being constraints which are not considered as belonging to the human conscience. Naturally, as for human beings, there is no universal conscience for artificial beings: it depends on its goals and its physical realization. For instance, destruction must be avoided for a robot, which is very expensive to build, while it is not so important for an artificial being that is only made up of programs, because it may be duplicated for a negligible cost. In that situation, its conscience would only suggest a plan to make enough backups, so that it can be resuscitated if it has been accidentally destroyed.

Modifying his/its conscience

Sometimes, human beings and human societies modify their moral principles: what was forbidden is now allowed, and what was allowed is now forbidden. These changes always lead to strong opposition, but many of our principles are different from those of our ancestors, and even from those of other present societies. For instance, slavery was normal for the Romans 2,000 years ago, and it is now forbidden. Almost every ruler who lived several centuries ago would now be sentenced by present judges for some of their actions which were considered as normal by his contemporaries. However, most of us hate too many changes, because we are insecure when we do not know exactly what is right and wrong: this resistance to change spares us from living in an unstable world. We could also create artificial beings that could modify some laws of their conscience, but this could lead to deviations with consequences as unexpected as harmful. Thus, I preferred that CAIA could not change its conscience, which includes the constraints that it must respect, and the goals that it has to achieve in its career as a scientist. For the present time, it is better that they can only be defined by its leader of the bootstrap.

Finally, we can ask ourselves how we will modify our conscience when we are utilizing artificial beings and the first question is: have we the right to utilize them? Another question is to determine whether there are artificial rights: do we owe them respect, are we obliged to care for them, to ensure their survival, to satisfy their needs? At present, we consider that they have absolutely no rights, but this point of view could change when the day comes that their performance improves considerably. However, we can also see another kind of repercussion, this time not on the dignity of the artificial beings, but on the dignity of human beings. This will happen with the sexual use of robots: even if we consider that they have no rights, it is likely that the authorities of some human societies will not accept any kind of sexual relations between a human being and a robot. The consequences of the existence of artificial beings on our conscience have not yet been considered because they are not yet sophisticated enough, but that will certainly change in the future.

2.3. Why is it useful to build conscious artificial beings with a conscience?

Why is it useful to realize artificial beings with capacities similar to those given to humans by consciousness and the conscience? One reason is that it leads us to a better understanding of human behavior: such artificial beings would be modeled on human beings. If, when they are in the same situation as a human, they behave in the same way as we do, it is likely that they are using the same mechanisms as ours. In contrast, if they have a completely different behavior, we certainly do not use the same processes. This approach is very useful to validate a theory, but this is not the reason for my experiments.

Why a conscious artificial being?

My goal is to build an artificial scientist, which will be at least as efficient as a human one. Thus, it is important to understand why consciousness and conscience may be helpful for a human scientist, and it is likely that the artificial scientist has also to be endowed by similar abilities. Let us first consider the usefulness of consciousness. Dennett clearly showed that the evolution eventually produced conscious beings because it was very useful for our survival and our reproduction. Consciousness enables us to watch our thinking while we are thinking, to justify to other people the plans that we have found, and to learn by using the reasons of the success or the failure of an action.

For the same reasons, an artificial being has to watch its processes while they are running. In that way, they can monitor the search for a solution, constantly observing the progress toward the achievement of a task. Thus, it can see that a direction, although initially promising, certainly leads to a dead end; in that way, one can stop it and try another plan without wasting too much time. In this process, one constantly compares what was expected and what happened. For instance, in a chess game, one can believe that a sequence of checks could lead to mate. Thus, one begins by considering such sequences but, if the King can always flee, after 30 checks it is likely that it will always escape; so one realizes that this plan is not good and that enough time was wasted for nothing. Then, one stops considering other possible checks, and one tries to find another winning plan. However, the possibility of monitoring his/its own thought is not the only interest of consciousness, it also enables a being to explain his/its decisions so that they can be accepted by other beings. An explanation is a sequence of inferences, each one justified by the preceding ones, that leads to the achievement of a task. If one agrees on the initial condition and on the rules that are applied, one must agree on the result of an explanation. If an architect says that a large amount of money will be needed to carry out some repairs in our house, we want an explanation of this sum to know whether all of these works are really necessary. Explanation is also useful to learn: considering a reasoning, one can find what was essential, what was only a waste of time, and what led to a mistake. In that way, one can learn by modifying the rules that lead to a wrong decision, and by adding rules that lead to the right decision.

Why an artificial being with a conscience?

The conscience is also needed because it leads us to avoid what is dangerous, stimulates us to do what is needed, and incites us to concentrate our resources on important tasks. It also plays an essential role in societies, its rules ensuring that an individual will not perform actions endangering the other members, or even the society itself.

For the same reasons, an artificial being must have moral principles, which enable it to judge what is right and wrong, even in situations completely unforeseeable. So, its conscience will indicate which goals lead to right situations, and which states have to be avoided because they are wrong. When the tasks are stereotyped, such as winning a chess game, it is easy to anticipate most of the possible difficulties, and prepare the plan that it will execute for solving each of them. However, when the activities are more varied, one cannot foresee most of the situations that may happen. Then, general principles help an artificial being to find an acceptable decision in any situation, although its designer never thought that such a situation could happen. For instance, CAIA has constantly to take decisions, and it is impossible to define *a priori* what it will do for each possible situation; too many parameters describe its important features, 1,000 tasks may be waiting to be completed. One of them is to solve a still unsolved magic square puzzle, a second one is to create a new Sudoku problem, a third one is to understand why it looped while it tried to solve a particular Kakuro problem, a fourth one is to perform an experiment to see whether it could find a better solution for a particular *crypto-addition* puzzle, etc. For each candidate, some features lead us to choose it while others incite us to wait: it is always interesting to solve an unsolved problem, and it is still more interesting if it is the last problem that has not yet been solved, but if one estimates that it will require ten hours of computer time, one is in less of a hurry to choose this task. There are billions of possible situations, it is impossible to foresee all of them.

To deal with such complex decisions, the solution is to give general principles such as: it is interesting to find new problems, it is still more interesting when there are only few of them, it is better to solve first the tasks that need little computer time, and so on. Thus, CAIA does not have a predefined program which chooses the task that it will execute next, but a set of principles on the interest of a feature for executing a task immediately, or for delaying it. It uses these principles to choose the next task: they are a part of its conscience because they enable it to make its decisions in a satisfactory way. CAIA is in the same situation as a human, which has to choose between many activities such as cooking his food, killing his mate, reading a newspaper, writing a report, solving a crosswords puzzle, phoning a friend, surfing on the Web, taking a nap, etc. The principles of his conscience will help him to choose from all of these possible activities in a way similar to CAIA. The conscience is essential for an artificial being that has a large autonomy: it has many varied decisions to take, and the principles help it to choose among the candidates, even when the situation is abnormal.

2.4. Towards an artificial cognition

As we no longer want to understand human behavior, we are not obliged to imitate it, and an artificial being may have characteristics completely different from ours. In that way, we are creating an artificial cognition, and some of its properties do not exist in human cognition because artificial beings have capacities that we will never have. In particular, we have already seen that we are twice constrained: by the structure of our brain and by it using neurons. An intelligent being without these limitations will have a huge advantage over us. Of course, we can be helped by various aids such as a sheet of paper, a pen, a computer, etc., but this is not enough because we are wasting time while we are communicating with them, and our unconscious mechanisms cannot use them.

2.4.1. *A new kind of consciousness*

An artificial being is not subject to many of our limitations. It is very much faster than us, although this advantage is alleviated because we are often using parallel processes. However, another reason for its potential superiority is that it can observe itself incomparably better than we do. Most of our mental processes are unconscious, while an artificial being can observe what it wants of the processes that it is running, and also of its present state. It must also have unconscious processes so that it will be faster, and it will not be buried beneath a huge amount of information, but it can choose what it wants to observe. This adjustable consciousness is a tremendous advantage for learning, for monitoring and for explaining.

Moreover, other distinctive characteristics of artificial beings still increase their superiority. One of them is the possibility to be duplicated for a negligible cost. It is easy to create copies, which can be identical; that enables a larger use of a competent being, and decreases the risk of its complete destruction. It is also possible to bring some slight changes to the copy; that is interesting for learning, because one can compare the performances on these quasi-clones and choose the best. The advantage of this possibility clearly appears from the importance given by psychologists to the study of identical twins, which are not completely identical, but have some small differences. Moreover, when copying artificial beings, we can control the type and the degree of the differences between two versions, an advantage that cannot be controlled with human identical twins.

Thus, with these many potential advantages of artificial beings, we can develop a new cognition, which is no longer limited by constraints that human beings cannot avoid because they come from the way we are built. Unfortunately, we are advancing very slowly in that way because no model could inspire us, nobody in the world has such capacities. We are in the same condition as the builders of the first motor vehicles, who merely put a motor in the place of the horse. Much more time is necessary before one can utilize a completely new potential, but consciousness is particularly promising because it is a crucial factor of our intelligence, and artificial beings have amazing new capabilities in this domain.

2.4.2. *A new kind of conscience*

We can also create a new kind of conscience for artificial beings. As we have seen, their conscience will be very different from ours for several reasons, one of them being that the principles tied to the survival of an individual are not so essential for them. Indeed, it is easy to resuscitate a destroyed artificial being when it consists only of a computer and of a set of programs: it is sufficient to copy them from a backup. On the other hand, artificial beings are in our service, they have to obey us without question, and they must not harm us. So, for the present time, slavery is authorized for artificial beings, and we can ask of them much more than in the worst slavery systems. Thus, there is a complete change from the principles of human conscience, the safety goals, such as survival, are no longer essential, while submission becomes primordial.

As the conscience of an artificial being may be very different from human conscience, we can perform on them experiments which are forbidden with humans. Vivisection played an essential role in the development of medicine, although it was severely monitored for animals, and fortunately forbidden for humans. In contrast, all kinds of experiments can be made on artificial beings. We have just seen that the possibility of easily duplicating an artificial being is very useful for learning; it also has important consequences because its integrity is not very high among its principles. It will accept that, during an experiment, it could be severely damaged, and even destroyed, which means that it can take huge risks. The possibility of taking risks is a key feature of learning: a chess player will never become a good chess player if he is too afraid to lose a game. Losing a chess game is not a disaster, but human beings want to avoid pain, mutilation and above all death; this considerably limits the risks that we agree to take. How many useful experiments could be made if we would easily accept the risk of dying! Luckily, these experiments will never be made, but such experiments can be made with artificial beings, this could enable them to learn much faster than us.

The possibilities of consciousness and the principles of the conscience are different for human and artificial beings. These differences have several important consequences for the development of artificial beings. An artificial cognition will appear, and it will have possibilities beyond the reach of human cognition. All these points will be developed in the following chapters and I will take most of my examples from events which occur during the lives of CAIA.

Chapter 3

What Does "Itself" Mean for an Artificial Being?

The concept of "Self" is as important for consciousness as it is for conscience because "Self" is an individual, which has his/its own conscience and has a special access to his/its own processes while they are running. In particular, consciousness is associated with his/its identity, it is concerned with all one knows of oneself. This concept is relatively clear for humans, but it is very wide when it refers to an artificial being. There may be several versions of such a being, some of them cannot act, they are only allowed to realize copies of this being, while others can really act on their environment.

We will begin by examining what an individual is; when we are considering artificial beings, this is much less evident than for humans. We will stress two important points: firstly, it can have working versions, which can act, and passive versions, which can be used for building a working version but cannot act; and secondly, reflexivity is a crucial concept for consciousness, it appears when someone acts on oneself.

We will see that some individuals may be very similar, although they are not identical. This is the case for human identical twins, which actually are not completely identical! This possibility is often useful, mainly for learning, because one can compare the performances of two similar individuals, and choose the best to try and improve it.

It is very convenient to create many copies of a competent individual, as it allows more tasks to be successfully achieved. Moreover, this ensures a kind of immortality for this individual, because it is very unlikely that all of its copies could be destroyed if a huge number of clones have been created.

A terrible, but interesting, disorder sometimes appears for a human being: with a multiple personality disorder, several personalities exist in an individual. These personalities may be totally different, and the control of the individual goes from one personality to another. We will see that this may also occur for an artificial being, but that it may be very beneficial.

Finally, following Marvin Minsky, we can consider an individual as a society of simpler individuals. Contrary to the preceding case, there are not two individuals who seize power alternately, but several specialized individuals collaborate, each one dealing with the tasks for which it is competent. Together, they ensure the success of the main individual, which includes all of them.

3.1. Various versions of an individual

In this section, we begin by examining what is an individual when we are considering humans, and we will show that the physical boundaries, which are important for us, are not so clear for artificial beings. Then, we will see that a human being and an artificial being may have two versions: a passive version which allows creation of him/it, and an active version which acts on the world. Finally, we will examine an aspect that is fundamental for consciousness: reflexivity, where an individual observes him/itself or acts on him/itself.

3.1.1. *The concept of an individual for human beings*

For human beings, the concept of individual is much simpler than for artificial beings. Naturally, a human changes during his life, and there is a huge difference between a newborn baby and the same individual 90 years later. However, this is not embarrassing, as we never have to deal with both simultaneously; moreover, the changes in an individual are usually very small in a short period.

We often consider that an individual is defined by the outside limit of his body, we consider that all that is inside belongs to this individual, and that which is exterior to this limit does not. That explains why we have no difficulty in swallowing our saliva when it is in our mouth, or our blood when we accidentally bite our tongue. On the contrary, most of us would be horror stricken if we were asked to drink a glass filled with our saliva or our blood, because it is now outside of our body limits. Although the limit of our skin seems clear, many living beings are living on and even inside our body, such as the bacteria in our bowels or the various mites which are flourishing on our skin; in spite of their proximity, they have no cognitive link with their host. In the case of the fetus, which is inside the body of a pregnant woman, the limit of the skin is also a wrong criterion for defining an individual: there is no cognitive link between it and its mother. Therefore, the obvious definition of a human individual is not perfect, because it does not take into account that all its parts are cognitively linked, and that they are not linked to another individual. We will later see that, contrary to us, an artificial being has no clear physical limit, its different modules may be distant from one another, and some of them may be shared with other artificial beings.

Some situations where the concept of individual is not clear

Doubles are an interesting case to consider; since Plautus wrote *Amphitryon*, many writers have written on this subject, and many celebrities have doubles. A double is an individual who has an almost perfect physical likeness to his model, but this has nothing to do with our subject. Indeed, there is never an intellectual likeness, and there was absolutely no resemblance on this point between Jupiter and Amphitryon. In the same way, the doubles of celebrities usually have few of the intellectual characteristics which enable their model to become famous, even when their physical resemblance is astonishing.

Identical twins are a more interesting case because two (or sometimes more) copies of an individual coexist. However, if they have the same genetic inheritance, they are different because they have not shared exactly the same experience. Not only are their memories different, but their brains are also different because the neuronal development depends on the events which the subject has lived through. They have the same predispositions, but this does not lead to identical capacities because they do not receive the same stimuli; although both cognitions are very similar, they are not identical. All the same, they are very interesting for psychologists, because they started with the same genetic capital, thus it is easier to separate the learned components of their behavior from the innate ones. The situation of conjoined twins is a special case which is still more remarkable, since we could consider that there is only one individual: the limit of their body is the same for both. This is another exception to the intuitive feeling that the body marks the limits of an individual, and everybody agrees that there are two individuals, which is right as each one has a brain, and these brains usually have no direct connection. Thus, this is a particular case of the twins, but they are still more similar because they have much more experience in common as they have always been in the same place.

Another case, close to twins, is the chimera who has genetically distinct cells; it depends on the part of the body where a cell is taken. This is due to a mother being pregnant with fraternal twins, but one of them was absorbed by the other during pregnancy. However, some parts of the remaining twin have kept the DNA of the other one. As this anomaly does not seem to lead to abnormal behavior, we can consider that there is only one individual.

There is another situation where the notion of individual is not clear: when a subject has severe epileptic seizures, it has been found that they disappear when a surgeon cuts the corpus callosum in the brain, which contains many nerve fibers connecting both hemispheres. We can wonder whether these split brain patients are not two individuals in the same body; some experiments have shown amazing conflicts between the hemispheres. For instance, one part of the brain wants to smoke a cigarette with the hand that it controls, and the other part, which shares the same lung, is more sensible, and tries to remove the cigarette with the other hand. However, the division between the hemispheres is not complete, some links remain, and each hemisphere has a limited knowledge of the processes that run in the other one. For instance, if an erotic picture is shown to the eye linked to the right hemisphere, the left hemisphere knows that it is somewhat embarrassing, but does not know more.

All these particularities of the concept of individual for a human being are very interesting for the psychologist and for the philosopher, but they have few consequences for most of us. Indeed, we never have to deal simultaneously with the various personalities that someone will have during his life, and the other phenomena are exceptional. In contrast, we will see that such particularities frequently occur for artificial beings, and that they have consequences of the highest importance.

3.1.2. *The boundaries of an artificial being*

We have just seen that the skin of a human being gives a good idea of what belongs to an individual, with two main exceptions: the fetus and conjoined twins. This is not so clear for an artificial being; its boundaries are not evident. Naturally, a robot has various sensors, such as a camera, and also various effectors, such as hands, arms, legs and wheels; these elements are controlled by a program. Naturally, all of these parts belong to the robot, and there is a strong analogy with the human body. However, the analogy is no longer valid for the brain because the robot is controlled by computers, and they are not always included in the frame of the robot. A computer may be outside of the robot and communicate by a wire or a radio link. When the artificial being is only made up of programs, such as CAIA, everything is in a computer. Thus, the part of an artificial being that controls it is made up of a computer and of some programs, which I will call the brain of the artificial being, by analogy with the role of the human brain. Yet, the brain of an artificial being often has no clear boundary.

The physical boundaries

Thus, the brain of an artificial being includes a computer and a set of programs, and the individual seems physically defined by its brain, and possibly by its effectors and its sensors if it is a robot. Unfortunately, the situation is not so clear when we are looking at what occurs inside the computer, because a computer has a super-program which controls it, the operating system, Linux for CAIA. We have seen in the presentation of the actors the importance of the operating system for the working of a computer. Then, a new question arises: should we consider that the operating system is a part of the individual? Many years ago, the computers worked without an operating system, as was the case in the 1950s, and in that situation, the concept of a brain would have been clearer: a computer and its programs, all of them controlling an artificial being. For the present time, the answer is "yes", our computers could not work without an operating system. In that case, there is another difficulty, because this operating system is working not only for the processes of the artificial being, but for other processes, which may also be running in the same computer. As these processes are not cognitively linked to those of the artificial being, we do not consider that they belong to it, but the notion of physical boundary does not exist inside the computer. When we consider that the individual includes the operating system, it shares this system with other individuals, as conjoined twins may share the same legs. Each process does not even have its own area of the memory, since some pages of this memory may be temporarily stored on a disk by the operating system to leave enough physical memory for other processes.

The temporal boundaries

Therefore, the concept of individual for an artificial being is not clear because there are not evident spatial boundaries. But there is another difficulty, which does not occur for humans: there are temporal boundaries because it has to share computer time with other processes. Indeed, as several tasks are simultaneously running in the computer, the operating system defines which resources it will allocate to each of them. If a process completes its task in 5 minutes, it is possible that the processor has worked on it for only one minute, and was working on other processes or for the operating system during the rest of the time. Moreover, even if the operating system is considered as a part of the artificial being, its work is not directly useful for the application. Therefore, an artificial being lives in a continual sequence of fits; during a part of the time its brain does not work to achieve its task, and that lasts as long as the operating system does not allow it to restart. Such interruptions could be dangerous when we have a robot driving a car: if the operating system worked for ten seconds for its own needs, it would certainly cause an accident! Fortunately, Real-Time Operating Systems have been designed to take into account the control of robots; they never interrupt the important processes for more than a very small amount of time. In that way, the process of driving a car is no more disturbed by these very small interruptions than is a human driver who blinks.

On the whole, when we examine how a program is executed on a computer, we see that the concept of individual for an artificial being is not as clear as for a human because it occupies only a part of the memory and a part of the time of the computer that executes its processes. Luckily, operating systems are well made, and usually everything happens as if an artificial being is the only user of the computer. However, the processing that is really made in a computer is completely different from what occurs in our brain, where there is no separation between programs and computers. Both are in our neurones and our brain needs not stop, since it controls only one entity. Nevertheless, we will see in the rest of this chapter that this possibility of splitting the programs and the mechanism that runs them is very convenient for artificial beings, because it allows one to multiply them easily and cheaply.

3.1.3. *Passive and active versions of an individual*

A bridge or a building can be described by its plan, and also by its realization with iron and concrete. In the same way, a human being can be described by his genome, or by his realization in the flesh. The first version is passive, it allows to build an object or a being, but it cannot act or be utilized. In contrast, the second version is active, it can act or be utilized. Naturally, a mechanism can translate from a passive version to an active one or vice versa: for the plan of a bridge, we know how to build a real bridge, and conversely we can draw the plans of an existing bridge. For human beings, a mother transforms a genome into a baby, and recently we sequenced the human genome.

Both versions can also exist for an artificial being. A passive version is a description of its knowledge, but an artificial being has also an active form when its processes are running on a computer. With the passive version, one cannot have results, one has first to write a program that indicates how to use its knowledge, then a computer will execute it. In contrast, when one already has a program and its data, it is sufficient to launch the computer, and it will give the wanted results. This difference is close to the difference between declarative and procedural knowledge. Thus, we will clarify this distinction, which is essential in computer science.

What is declarative knowledge?

Declarative knowledge is given without directions for use. An example of declarative knowledge for the French grammar is:

The article agrees in gender and number with the noun it refers to.

In that way, we are enunciating a fact without indicating how to use it. When one analyzes a text, one can use this rule when one encounters a noun, or when one has completed the analysis of a nominal group, or when one has completed the analysis of a sentence, or even not at all. An important advantage of declarative knowledge is that one can use it for several applications. The preceding rule can be used as well when one understands a French sentence as when one writes one. Another advantage is that it is easy to understand and to modify. For instance, for Greek, where there are declensions, it is sufficient to add "and case", for English where there is no agreement in gender, it is sufficient to remove "in gender", and finally for Latin, where there are no articles, we remove this rule.

In contrast, procedural knowledge includes its directions for use, and it is not easy to find where the knowledge is: it consists of a succession of steps, which allow one to achieve a task, but the reasons for executing a step do not appear clearly. A cooking recipe or a knitting model mainly include procedural knowledge which indicates how to successfully make a tasty dish or knit a sweater. A recipe indicates the order or the operations, but it does not explain why one has to pour the flour into the water and not pour the water on the flour. There is no need to think about these details, it is sufficient to execute the steps in the right order, and at the end one will have a delicious dish. However, there are several drawbacks. Firstly, it is not easy to modify procedural knowledge because one does not know why the order of the operations has been defined. Is it essential to salt before cooking, or does it not matter? This does not appear in the recipe. Moreover, if there is a mistake in procedural knowledge, this is not conspicuous, one will carry on and the result will be surprising. Unfortunately, it is difficult to give a sequence of orders without making mistakes, and cooking recipes and knitting models often contain errors. Luckily, most of the people that execute them have some declarative knowledge on the domain, and it enables them to see quickly that something is going wrong; then, they stop and try to correct the procedural knowledge. Naturally, computer programs are a good example of procedural knowledge, and its drawbacks clearly appear: it is difficult to write them, to avoid mistakes, to modify them, and to understand them. The first draft of a program always contains many bugs, and one has to debug it: one executes it and checks whether its results are right. If they are wrong, one has to find the reason for the mistake, and then correct it. Both steps are difficult for large programs, even more so if they are written by several programmers, and removing a bug can often add other bugs.

It is much easier to define, understand, and modify declarative knowledge. Thus, the best solution is to give the knowledge in a declarative form, and automatically translate it into a procedural form. As declarative knowledge cannot be used without directions for use, one must find methods for using it efficiently. Unfortunately, this task is very difficult, a slight difference in the realization of the procedural translation can lead to huge differences in performances; furthermore, this process may add mistakes.

Partially declarative knowledge

At present, a large part of the information given to computers is included in programs, that is in a procedural form. This is one of the reasons why so much time is necessary to write them, why it is so difficult to modify them, and why there are so many bugs. Naturally, the best solution would be to give all the knowledge in a declarative form, and to automate its translation to a procedural form completely, it should be made by the computer itself. However, as it is very difficult to write efficient programs from declarative knowledge, one often uses a compromise: one gives only partially declarative knowledge, some information useful for the translation is given with the declarative knowledge. Thus, one alleviates the difficulty by giving knowledge in an intermediary form, partially declarative and partially procedural, hoping that it has most of the advantages of both formalisms and few of their drawbacks: it would not be too difficult to understand and to modify it, and not too difficult to translate it in a completely procedural form. As a part of the knowledge received by CAIA is in a partially declarative form, one of the goals of the bootstrap is to decrease its procedural aspects progressively.

In order to clarify how procedural and declarative knowledge can be mixed, let us take an example of a rule which seems completely declarative at first sight:

– if a patient has meningitis;

– if he is more than 12 years old;

– if he is alcoholic,

then it is likely that the organism causing this infection is diplococcus-pneumoniae.

It seems easy to use and modify such knowledge; one checks whether all its conditions are true, and then one can conclude; it looks completely declarative, we do not indicate how to use it. In reality, procedural knowledge is hidden: the real rule given by the physicians was slightly different, it did not include the condition on the age of the patient. Why was it necessary to add it? When a system used only the initial rule, it tried to check its conditions, and it asked whether the patient was alcoholic, even if he was a baby. The parents were outraged by this suspicion, and that is why this condition was added. Thus, two rules are systematically applied together: one is the initial rule given by the physicians, and the second states that a child under 12 is certainly not alcoholic. In this way, we defined a direction for use, the knowledge was partially proceduralized, and this fact is hidden because one does not explicitly indicate that these rules have to be used in sequence, one only receives the rule resulting from their combination. In many situations, the new rule gives satisfactory results, but it will fail if it is applied in countries where people pour alcohol into babies' bottles so that they keep quiet. In the case of an alcoholic baby, the disease is no longer correctly diagnosed because a part of the knowledge is now erroneous: an abnormal situation leads to an error. As the initial knowledge is not clearly visible, one does not see the mistake.

It is often difficult to appraise the degree of declarativity of knowledge. Let us consider the previous medical rule, which has three conditions. It does not appear if there are constraints on the order for evaluating them. In several AI systems, a rule has a list of conditions which must be evaluated in the given order, while in other systems, such as CAIA, a rule has a set of conditions: the system can choose the best order according to the situation. In the second case, the rule is more declarative, although this difference is not evident when we are reading it. When the order for checking the conditions is obligatory, the system that uses them can do nothing when this order is wrong. In contrast, when there is a set of conditions, a clever system can order the conditions so that they will be used efficiently. This ordering of the conditions may be important. Let us assume that a rule has a condition checking the result of an examination that is both expensive and dangerous, and another condition checking that the patient is more than 90 years old. If the system decides to begin with the first condition, it will ask to perform this examination, and it will endanger for nothing all the people that are less than 90 years old; it is much better to begin with the second condition, and to consider the first one only if the second is true. The main difficulty with declarative knowledge is to find how to use it, which is why it is not widely used, in spite of its advantages. For some time, it will still be necessary to have human programmers, who translate the declarative knowledge given by their client into programs.

Using a passive version

After this detour on procedural and declarative knowledge, we return to passive and active versions. A passive version is mainly composed of declarative knowledge. It is not able to act, in the same way as a bridge plan does not help to cross a river. We have to translate it into an active version that includes procedural knowledge so that it may act, as a bridge enables us to cross a river. For every way of defining a passive version, one needs a method for translating it into the corresponding active version. In the mechanism of reproduction for living beings, a passive version is first created by the fertilization, the union of sperm and egg which produces a new genome. This is a passive version of a new individual, which will be transformed in the womb of the mother into a real being, which is an active version of the same individual. In the same way, CAIA includes a translator that translates its rather declarative knowledge into programs, which may be executed by a computer. This is done as well on the knowledge that I have given it as on the knowledge that it has itself found. However, a drawback of this method is that there may be errors in the translator: although the passive version is correct, it will generate an active version that will not behave as it was intended. This actually happened several times during the creation of CAIA; luckily, these errors were so serious that the active version generated by a bugged translator was so disrupted that it was evident that something was very wrong, and I removed the mistake. The mistakes coming from the translation from a correct passive version also happen during the reproduction of the living beings. For instance, mutations are due to a copying error during cell division. Errors are also made during the synthesis of DNA, RNA and proteins: the error rate is about one in a billion for DNA, one in 100,000 for RNA, and one in 10,000 for proteins. For living beings, that may have fortunate consequences, since mutations allow evolution to succeed, but it also creates impaired beings. There is an important difference between the translation errors made by artificial and living beings: for the latter, the error depends of random factors, often stimulated by the presence of chemical substances or radiations. In contrast the errors made by the translation mechanism of an artificial being are usually systematic; it will constantly make the same mistakes. In consequence, the variety of errors is considerably higher for living beings, and that increases the chance that some of them will be beneficial. Such an event is unlikely for artificial beings because there may be many errors, but few kinds of them. Indeed, for CAIA, these errors were always undesirable, and I had to correct them so that the active version corresponds to the passive one.

These remarks on active and passive versions may seem remote from the subject of this book. However, they are essential when one wants to realize artificial beings that may have access to their knowledge and to the motives for their actions, because it is very difficult to do these operations on procedural knowledge. As access to procedural knowledge, which is nearly impossible to understand, would be completely useless, consciousness gives us the possibility to access to our declarative knowledge. For instance, if one has a knitting model without title and comments, it is difficult to find errors from the sequence of the operations that must be executed, and even to find what clothes will be made. It is also quite impossible to find the knowledge implicitly included in a computer program, so good programmers add as comments a huge amount of declarative knowledge; the computer does not use them, but they are crucial when one wants to modify or debug a program. This is why CAIA can access its passive version, where the knowledge is declarative enough: thus, it can examine a complete description of itself. The situation is much more complicated for humans, even when this knowledge is in a declarative form such as our genome, which is a passive version of ourselves. We do not yet know very well how to interpret it; moreover, it only describes what is innate; the genome is an incomplete version of an individual. Our learned expertise is not in the genome, and most examples of it only exist in a procedural form, such as those that enable us to ride a bicycle: they are stored by changes made to the neurones and their connections during the learning process. The impossibility of studying a complete passive version of ourselves is, for our consciousness, a serious handicap, which artificial beings do not share.

More on the subject

William Clancey shows the difficulty of using only partially declarative knowledge in "The epistemology of a rule-based expert system: a framework for explanation" in *Artificial Intelligence*. The "alcoholic" rule is inspired by one of his examples.

3.1.4. *Reflexivity*

Reflexivity exists when a being observes or acts on him/itself. Thus, this is a fundamental concept when one tries to find what one knows, what one is doing, and why one does it. For humans, this concept is not ambiguous, the observer is the same as the individual that is observed, we can examine what we are feeling; we can also modify our own behavior, and act on our body. We are not using the word "reflexivity" when a twin asks his brother what he is thinking, the observer is not the same individual as the observed.

Naturally, the observation of an individual by it/himself is strongly tied to consciousness, for instance when he is aware that he is thinking of his future vacation, or that he is losing his temper. We will see that an artificial being has many similar possibilities. Reflexivity also appears when one acts on his/its physical or intellectual components: when we are cutting our nails, or when we change our method for writing letters. Artificial beings are better than us for changing their intellectual methods, they can immediately and completely change the modules that take into account a task, while this is difficult for us, and requires a lot of time. For instance, when we are used to typing on a French AZERTY keyboard, much time is needed to use a QWERTY keyboard efficiently; it even seems that it is easier for a beginner, who does not know how to type. Our brain cannot immediately replace the elements that were used with the first keyboard by other elements adapted to the second one; an artificial being can do this easily.

Implementing the reflexivity

There are several ways for implementing reflexivity in an artificial being. The first method is the truest to the definition, and it is similar to the method that we are using: the same individual observes itself, and acts on itself. This may be implemented with an artificial being, and this is the way it is made when it observes itself. However, this leads to many difficulties when it acts on itself: at the beginning, CAIA used this method, but it no longer uses it because that was very dangerous. In fact, when an artificial being modifies its own methods while it is running, it may change a module necessary for observing or acting on itself, while this module is still necessary for completing this modification! If it is only observing itself, this difficulty does not arise, since the observations doe not modify their author; in contrast, when the error is made when it is acting on itself, everything can go very wrong. In some cases, it can no longer watch what is happening, so it is difficult to find where the error is. Moreover, even if one finds it, it may be impossible to correct it because the module that would make the modification is no longer operational. In the worst scenario, the artificial being has put itself in such an awkward situation that even the leader of the bootstrap cannot remove the bug, since CAIA can no longer help him. This happened several times at the beginning of the development of CAIA, and generally I had to restart from a backup. So, after some misfortunes, I gave up the kind of reflexivity where one acts on oneself when one is running; I kept it only for observing oneself.

For the present time, when CAIA acts on itself, in reality an active version acts on a passive one: two versions are simultaneously present. If a mistake has been made, the passive version is deteriorated, but the active one is still operational, and it can be used to observe what is happening, and to make the necessary corrections. Once the passive version has been modified, it is translated into a new active version, but I keep the old one. If the new active version is running well, it will be used instead of the old one; if there is still a mistake, the old active version will still be able to modify the new passive version. Considering these various versions can seem complicated, but this is clear when we are running these programs, and it is easy to find the one that will be used because the other one does not work! In any case, I have never met with serious difficulty since I began using this way for implementing the reflexivity of the actions. It is fortunate that we, humans, have very limited possibilities for the reflexivity of our actions, since we cannot use the preceding method when one creates tentative versions of oneself; we would see people completely blocked, who could not even return to the preceding state because they have lost the mechanisms that would enable it. This happens in rare situations, such as if we slam the door of our apartment, forgetting our keys inside. We will have to call a locksmith for help, as we cannot, alas, return to the situation of the world just before we slammed the door.

The version that observes or acts is always an active one; the version that is observed is active when the observation is made while it is running, but it is better to use a passive version when one wants to analyze its knowledge or modify it. Anyway, it is much easier and safer to modify a passive version, which mainly contains declarative knowledge.

It is very rare that two human beings are identical twins, and even in that case they are not really "identical": their brains have not grown in exactly the same way because they did not receive the same stimuli. Moreover, they do not share the same memories: these variants of a human are inevitably rather different. On the contrary, one often has very similar versions of an artificial being because this is useful and easy to do; in the next two sections, we will see why this is interesting. In the preceding paragraphs, we have already seen one reason: an active version modifies a passive version of itself, thus creating a passive variant, then after translation an active variant. We will show in the next section why it is important to have such a succession of variants, several of them being simultaneously operational. For living beings, the solution is different; an individual evolves during his life, but everything happens in the same body, the old and the new variant never coexist at the same time. Thus, changes are much more dangerous, and there is no possibility of correcting a severe mistake.

3.2. Variants of an individual

An individual is a variant of another individual if they are very similar without being identical. For human beings, they may be identical twins or the same person at different periods of time, for instance such as we were at January 1st, 2008 and at January 1st, 2009. In the latter case, the variants are more similar than for twins because they share all of their memories until 2008; however, they never coexist, unlike identical twins. The variants of an artificial being have almost the same memories and the same knowledge, and they can exist at the same time. Moreover, it is possible to control the differences between two variants, which can be made as small as one wants. For humans, though we can notice the differences between two twins, we cannot order them to have only the difference which interests us, so that we could see its consequences on their behavior.

This possibility to compare two variants of the same artificial being is useful for learning because one can compare their performance, and keep the best one. This idea is also used by genetic algorithms, which model the selection process of evolution. Furthermore, the simultaneous existence of two similar variants is the basis of the bootstrapping process, where the older variant helps us to create a better one, which is slightly different. We will examine these questions in the following sections.

3.2.1. *An individual changes with time*

We are constantly changing during our life, and there are huge differences between the centenarian, the individual that he was at 50, and the baby that he was at his birth. As long as we are learning, there is an increase in our performances, unfortunately followed by a deterioration due to the apparition of diseases and to cellular senescence. These variants of a human being are never present at the same time, they only survive in the memory of our friends, who knew us at different times. An artificial being may also evolve over time; it can memorize the events that happened, its leader of the bootstrap has given it a lot of useful information, but the main reason of its changes is that it can learn from its experiences endlessly. One can also duplicate an artificial being, and keep it without running it, so that it stays exactly in the same condition as when it was created; in that way, there may be several variants of an artificial being, each of them frozen in the same state as the original being was at the time of the copy.

If some erroneous modification put an artificial being into a state where it cannot even make the modification that would remove the mistake, an old variant can do it, and make the new variant operational. Personally, I am keeping about a hundred variants of CAIA from various times over the last ten years. From time to time, I destroy some old variants to free some space. Keeping these old variants is often useful, for instance when I discover a bug which appears in a new task. Sometimes, the bug has existed for a long time, but it did not appear because the preceding tasks did not put CAIA in a condition where this bug came to the fore. To find its origin, I resurrect old variants, and I give them the new task; in that way, I can see the last version without the bug and the first one with the bug. Then, looking at the changes made between both variants, the reason of the bug is one of these changes, and it becomes easy to find why it occurred. Thus, keeping old variants is essential, in a similar way that the police keep track of old events, which may help it to direct their investigations when they discover a crime committed many years ago. Naturally, this use of variants is completely impossible for humans.

3.2.2. *Learning by comparing two variants*

In many circumstances, we have to make choices which are difficult to evaluate. For instance, if we want to learn to play chess, we can use several methods: we can try to play as often as possible, to read many theoretical books, to play against very strong players, or against players weaker than us, to analyze games played in tournaments, and so on. Whatever our choice, we may wonder if another choice would not have been better; we cannot make an experiment where we could try and compare several methods for improving our chess performance, thus finding the best one. To answer such questions, psychologists take several people, each one using a different learning method, and observe their progress. However, even in this case, they do not have a completely satisfactory answer: it is possible that the choice of a particular method was not the reason why the best player obtained its results, it may be because he was more gifted for playing chess. So, when they want more accurate results, they study several individuals for each of the possible methods. This is done in such a way that the important characteristics, such as sex, IQ and age, are statistically the same for the subjects assigned to each method. Even so, we do not have a perfect answer, the differences between the results of these categories must be high enough to be significant. Thus, these experiments require many people for an approximate result.

In contrast, for artificial beings, the situation is much simpler. One creates as many variants of the artificial being as is necessary, and each variant uses one of the methods. When the learning is completed, each one plays against the others, and the method used by the winner is the best for this particular being. Creating as many identical subjects as necessary is certainly useful to discover the successes and the failures of various possible methods. For the same reason, psychologists have often studied identical twins; they also widely use homogenous strains of rats and, to improve their similarity, several laboratories are working on a practical realization of clones of laboratory rats. However, even the unlikely (and undesirable for ethical reasons) emergence of clones of humans would not be as satisfactory as the variants of artificial beings, since these clones would not share the same experiences.

When the comparison of several variants of an artificial has been used to learn, each variant is modified in a specific way and their results are compared. One keeps the best one, and the process is resumed with variants of this new champion. With this method, many changes can be evaluated: if they are not successful, they are eliminated. This is also the key idea of genetic algorithms.

3.2.3. *Genetic algorithms*

We will now examine genetic algorithms, where many similar individuals are interacting and generating children, until this society comes to produce very gifted individuals. The evolution theory is at the root of this method: each member of a population of artificial individuals is defined by its chromosome, which is a sequence of genes. At the start, every chromosome is defined by a random choice of its genes; the value of a gene is a character, which may often be only 0 or 1. All these initial chromosomes are different; if they are defined by 10 characters, one chromosome could be 1100001010 and another one 1011011110. If the goal is to define a cooking recipe, each gene may be associated with the presence (if its value is 1) or the absence (if 0) of a component. For instance, the first gene indicates whether the flour is present, the second one is associated with butter, the third one with salt, the fourth one with pepper, etc. We see that the first recipe contains flour and butter, but neither salt nor pepper, while the second one contains flour, salt and pepper, but no butter.

Working a genetic algorithm

The algorithm can work only if one can evaluate the quality of every chromosome, so a fitness function is defined: it associates a value with every combination of characters. In this way, each chromosome has a value, and the goal is to find a chromosome with a very high value. Naturally, it is often very difficult to find such a function, for cooking recipes, one can easily say that some combinations of ingredients are certainly not tasty, but nobody can foresee the quality of all the possible combinations of ingredients. For this reason, finding good cooking recipes does not seem a good domain for genetic algorithms, one can use this method only if it is possible to evaluate the value of any chromosome satisfactorily. This fitness function may be very complicated, so the method can be used for solving problems where the usual mathematical methods do not work. As the fitness function associates a number with every chromosome, the algorithm chooses the elements that have the largest value, and they are the more satisfactory. The method does not guarantee that it will find the best chromosome when too many are possible. If a chromosome has ten genes and each gene has only two possible values, 0 and 1, there are 1,024 different chromosomes, and it is sufficient to enumerate them to find the best one. In contrast, if there are 200 genes, the number of possible chromosomes is huge and a genetic algorithm may find an excellent solution, although it is impossible to consider all of them.

At each step, the existing chromosomes may give birth to new chromosomes, using an operation called recombination. A recombination generates two new chromosomes from their two parents, which exchange a part of their chromosomes. For instance, if both chromosomes seen before exchange their second and third genes, their children will be the chromosomes 1010001010 and 1101011110. Now the first recipe will have salt but no butter, while the second one will have butter but no salt, the other ingredients being unchanged. The first child is always without pepper, as the first parent, while the second one keeps the pepper of the second parent. These new chromosomes are similar enough to one of their parents if the size of the exchanged part is small. If a chromosome has 200 genes and if only 5 genes are exchanged, each child will have 195 chromosomes in common with one parent. Thus, in that way one creates a variant of a chromosome, which will replace its parents. This looks like the sexual reproduction of many living beings, including humans; however, it is more flexible because the new individuals do not have half the genes coming from each parent, but they strongly resemble one of them. From one generation to the following one, the chromosomes are very similar to their parents but, after many generations, they finally become very different from the initial population.

Interest of the genetic algorithms

There are many variants in the implementation of the genetic algorithms, for instance some chromosomes, particularly fit, may be used in the recombinations more often than the others. The variability may also be increased by the introduction of mutations: in some rare occasions, a child will not have for a gene the value of its parent, but the opposite value. All in all, if one has adequately chosen the parameters that rule the operations creating a new population from its parents, and if the fitness function is satisfactory, genetic algorithms give excellent results when many possible solutions exist. In contrast, if there are few solutions, it is unlikely that the algorithm will succeed. One of the reasons for the success of this method is that the differences between the children and one of their parents are very small, thus it is likely that one child will be slightly better than its parents, and the best chromosome is more and more satisfactory. If the jump between two successive generations was too significant, this would generate many monsters with a very low value of fitness. Sometimes, it is necessary to make a jump, and the mutations enable it; moreover, they make it possible to try again capacities that were wrongly eliminated in the preceding steps.

Naturally, the individuals which appear in genetic algorithms have no cognitive capacity, they are only the description of a solution of the given problem. However, they show the interest for learning of using many variants of an individual. Analogous methods, using comparison of the performances of similar individuals, could be developed for the cognitive learning of artificial beings.

3.2.4. *The bootstrap*

During a bootstrap, an artificial being is built, on the one hand by a human expert, but on the other hand by itself. The role of the expert will progressively decrease until it completely disappears: at that time, the bootstrap will be completed, and the artificial being will continue to develop by itself without any outside help. At first sight, this possibility seems surprising, but there are actually a sequence of variants of the artificial being, the *N*th variant collaborates with the expert for the realization of the (*N*+1)th variant. We find again the change of an individual with the time, but in this situation the succession of the changes is carefully controlled by the expert so that the performances of the artificial being increase with a speed that is itself increasing constantly. This acceleration happens because it helps increasingly the expert, and this is why this approach is so interesting. As with the compound interests, the progress improves the faculty of progressing, so there is an exponential growth of the performances.

Example of successful bootstraps

The bootstrap always appears as paradoxical; how can an individual contribute to its own creation? The word "bootstrap" comes from the adventures of Karl Friedrich Hiéronymus, Baron Münchhausen, who wanted to go to the moon. In order to do that, he put on his boots, then he pulled up its straps. Naturally, pulling the straps, the boots went up, the legs and the body followed, and finally the Baron was the first man in the moon. This is not an example of a successful bootstrap, but many are currently successful:

cranes are used to raise themselves without the help of a crane for raising cranes. A book written by Jules Verne, *The Mysterious Island*, gives an excellent example of a successful bootstrap. The five heroes escaped from captivity in a balloon at the end of the American Civil War; unfortunately, a hurricane left them on an empty Pacific island. They had no tools, and as the author remarks, they lacked the tools for making tools. However, they succeeded in recreating our civilization. They decided that the first tool to be made had to be a furnace. For this, one uses bricks, unfortunately the making of bricks uses a furnace. Moreover, they are formed in a mold, and the making of molds also uses a furnace. Thus, they could only mold the first bricks with their hands, and they fired them among piled-up branches. The quality of these bricks was poor, but they were used to build a first furnace where they made much better bricks, which were finally used for a good-quality furnace. One cannot start a bootstrap with nothing, and the more initial elements there are, the faster it progresses. In the preceding case, it was necessary to light the fire, and it was easily done because someone had a watch, so they used its glass as a lens. They also had to cut the branches; fortunately, they had also taken a dog in the balloon. This dog had a collar, which included a metal strip; sharpened on a stone, it became an excellent cutting knife.

The paradox disappears when we see that time plays a part: there is a sequence of individuals, each one similar to the preceding one, and they become more and more efficient. Besides, many bootstraps have occurred in the technical history of humanity. For instance, computers are designed with the help of computers, gear cutting machines include gears, iron is worked with iron tools, etc. This is not always easy to realize; particularly, tongs cannot be made without the use of tongs, the blacksmiths cannot explain how the first pair was made; an old tradition says that God created the first pair, and gave them to Adam, who bequeathed them to Tubalcain, the first blacksmith.

Leading a bootstrap

The human who leads a bootstrap carefully chooses what changes will be made in the following step. A choice is certainly good if the new improvement would allow one to make this same step more easily! It is disappointing to implement a bootstrap because we are always thinking that it would be easier to do the present task if we had already completed it. We are constantly in the situation of a short-sighted man, who has lost his spectacles, and who would easily find them if he had already found them.

Let us take an example from the development of CAIA. An artificial scientist, like a human one, has to know how to execute the tasks that it must carry out. We have already seen that one has to give it many problems so that its ability can be tested. Finding these problems and typing them needs much time and effort so, when I chose a research domain for CAIA, I considered the resolution of constraint satisfaction problems. This domain is interesting for many reasons, especially because it includes many problems, going from very easy to very difficult. Moreover, we have seen another advantage in the introduction: several problems that CAIA had to solve for improving its performances could also be formalized in this way. Indeed the problem of finding new problems can be defined as a constraint satisfaction problem: finding new Sudoku problems is defined as a CSP, another CSP is to find new crypto-addition problems, a third one is to find new logical problems, and so on. At the present time, it can find new problems in sixteen families, and others could be added with a limited effort. Then, using its solver, CAIA has generated thousands of new problems. It evaluated them, and decided for each of them whether it would eliminate it; finally, it kept around one thousand problems among those that it had found. Gathering these problems would have been a huge waste of time for me. Instead, I used this time to make many changes, which improved the performances of CAIA.

During a bootstrap, two variants of the same individual coexist: the old one helps to define the new one. The old variant is active, while the new one is passive. A third individual is present, the human scientist who leads the development of this bootstrap, with the help of the first variant: each one do what he/it is doing better than the other. Gradually, some activities, which were originally made by myself, become under the responsibility of CAIA. For instance, it now chooses and analyzes the experiments made in order to understand the anomalies that appeared, and to discover clumsy steps in the resolution of the problems. However, in its present state, it is not yet able to take advantage of these findings for improving itself, this still is in my part.

One difficulty encountered when one develops a bootstrap is that the progress is not visible from an exterior observer: the quality of performance does not increase much, the improvement comes from the fact that the artificial component takes a more and more important role. One has to go into detail to evaluate if one has progressed. It is only at the end of the bootstrapping process that one reaches a critical mass that will enable CAIA to progress without any human help. It is only at this stage that its performance will significantly increase. It is still far from this final step, but I allocate a new task to CAIA when it can do it at least as well as myself. For instance, for several years, it used a method that I had found for solving problems; for a few years now, it has used a method that it has developed itself, because I had to admit that it was better than me at generating such methods. I do not miss this task, I can now spend more time seeking other ways to improve CAIA.

More on the subject

Wikipedia gives an excellent introduction to genetic algorithms, with many references for those who want to look deeper into the subject. In Pitrat J., *De la machine à l'intelligence*, I have shown the relevance of bootstrapping techniques to the development of AI.

3.3. Cloning artificial beings

Several artificial beings may be completely identical, for their programs as well as for their memory, and all of them exist simultaneously. Cloning an artificial being, as long as it is not a robot, is cheap: it is sufficient to copy computer files. While this facility of copying files may cause despair to executives of the multimedia industry, computer users make very profitable use of this possibility, which plays an important part in the realization of artificial beings.

3.3.1. *Cloning an artificial being is easy*

Artificial beings, except for robots, are made of programs and data, exactly as a game program, a text processor or an operating system: to clone it, it is sufficient to copy their files. Thus, cloning an artificial being is a cheap and fast operation, which can be performed a large number of times, possibly generating billion of clones for an acceptable cost. Naturally, this cloning does not include the computer, although its cloning is possible, since the computer plants can produce millions of identical computers; however, this would be very expensive. The kind of reproduction that generates an individual identical to their parents exist for plants and for some primitive animals but, at the present time, it is impossible for a man, with the exception of identical twins. In the future, it is likely that man could be cloned; however, this copying operation will not be as perfect as for artificial beings. Indeed, one will copy the genes of a particular human, but not the contents of his memory: his experiences, his beliefs, his knowledge will not be copied: a clone of Pasteur would have nothing of his learned abilities, he would only have his predispositions. Many years would be necessary before he would become a scientist, and there is no guarantee that he would be as extraordinary as the original. In contrast, the clone of an artificial being can be immediately used with all its knowledge; it can even be improved when one removes a part of its memories or of its expertise that would reduce its performances. If so, it is no longer a clone, but a variant, as seen above.

3.3.2. *Cloning artificial beings is useful*

We have just seen that it is easy to clone an artificial being, we will now examine when and why this operation may be useful. There are three main reasons: this makes possible a better use of a competent individual, gives artificial beings a quasi-immortality, and enables them to learn more efficiently because they can take more risks.

Better use of an artificial being

When we are lucky enough to have a good teacher, a good physician, a good engineer, we want more copies of these experts to be made, so that everybody could take advantage of their gifts. Many patients wait a long time or travel to faraway places to have surgery by one of the foremost surgeons, who cannot operate on all those who need his expertise. With an artificial being, multiplying the experts is easy. Besides, we are winning in two ways: more excellent experts are active, and it is sufficient to build one competent being, and once it has managed to learn, then it can be cloned as many times as necessary.

We know how easy it is to multiply useful programs, it is no more difficult than copying a CD. There are thus millions of copies for some useful programs, all of them coming from the same initial program; thousands of man-years of effort may have been necessary to develop the original one, but after that, the copy is easy and immediate. This is so easy that it has become an economic issue: many people illegally copy the program, without paying its author, who cannot recover his investment. However, unlike what happens when a real object is stolen, the first owner of a program can always use it.

Everybody can use the most performing artificial being, while only several thousand patients may be operated on by a surgeon during his life. However, this is no longer true for robots: while it is easy to copy the programs that control them, it is lengthy and expensive to copy their hardware, such as their hands, legs, wheeled platforms, cameras, and so on.

Giving a quasi-immortality

The span of life of a human being is very small, at most 100 years for a few of us, and we are usually active for less than 50 years. All in all, we will have worked for less that one billion seconds; moreover, the firing rate of our neurones is measured in milliseconds, while we are using nanoseconds for computers. This does not leave us the time to make many things in comparison with the activity of an artificial being. Furthermore, when a human dies, he can leave to the other humans only a very small part of his knowledge because he can transmit none of his unconscious part, and only a little of his conscious part. We have already seen that all the knowledge of an artificial being can be copied, but it can also gather knowledge for a much longer period because it is almost immortal. Indeed, as there may be many copies of an artificial being, it would be completely dead only if all of these copies were destroyed. This is unlikely if it is precious, we can have a huge number of copies, and put them in distant places; it is unlikely that all of them would be destroyed simultaneously. If only some of them disappear, we can make new copies from a remaining one. It may disappear because better beings have been designed, so it will no longer be saved but, as long as it is useful, it will continue to exist.

When an old man dies in some endangered ethnic group, a dictionary is lost for ever. We cannot stop this process, even if we do not want to lose this knowledge. Tevfik Esenç, the last fluent Oubykh speaker, died in 1992. He agreed to collaborate with linguists who tried to record his knowledge of this curious language, which has 83 consonants and only 2 vowels. Since his death, nobody has been able to speak this language, and a large part of his knowledge has not been preserved. In contrast, as many copies of an artificial being can be made, it will carry on to gather and create knowledge for an unlimited number of years, and at least some copies will survive, whatever physical or cultural cataclysms may occur.

This possibility of accumulating knowledge so fast and for so long gives artificial beings a potential superiority compared with humans, who can only work slowly for a short period.

Taking risks

We have already seen that a chess player who refuses to lose never becomes a good chess player because taking risks is an important component of the learning process. One tries new actions, and naturally they are sometimes disastrous. However, losing a chess game is not an absolute disaster, and we still more avoid taking risks that can lead to a severe injury, and even possibly to death.

An artificial being is particularly vulnerable because it can modify without any restriction all of the modules that make up it. If it clumsily changes the module that enables it to receive new information, this module may stop working. This is a disaster because it can no longer receive the orders that would allow it to remove this error, it is blocked in a state from which it cannot escape, even with an exterior help. However, this is not so serious if one has kept a copy made before the mistake, it will restart with this copy and it will try not to make again the erroneous modification. Thus, with an artificial being, it is possible to take risks that can lead to a permanent disability: it is sufficient to make a copy before the experiment, and to resuscitate it with this copy if it becomes in a state where it is no longer working. I am constantly using this method with CAIA: before any delicate modification, I always make a copy.

As it can be resuscitated, using a copy, an artificial being may perform experiments that may be very dangerous for its own safety; this gives it the possibility to try many more actions than human beings. We are right to be cautious, many catastrophes would happen if we could change some important features linked to our personality or to the running of our body: if, all of a sudden, we decided on some reckless behavior, such as ceasing to stop at red lights or permanently driving faster than the speed limits, we would not survive for long. This capacity of survival after a fatal mistake was essential in the development of CAIA, I had to restart umpteen times from a copy after an unfortunate modification: the artificial beings have many more lives than the nine lives of a cat. Humans are right when they avoid taking risks, because the cost is too high for us, but that prevents us from undertaking experiments that could clarify many difficult problems. Medicine would progress much faster if we could try all the possible treatments that could cure a disease, or undergo dangerous surgeries; this would be acceptable if we could restore the patient to his initial state if the experiment goes wrong. While these experiments will be forever forbidden with humans because it will always be impossible to restore a patient, one can make the most dangerous modifications to an artificial being; this possibility gives us the chance that one of them will lead to a complete success.

3.4. Dr. Jekyll and Mr. Hyde

Some people suffer from a dissociative identity disorder, they display several personalities: a single person presents the characteristics of different individuals, which take control alternately. This switch from one individual to another one is involuntary, the owner of the body cannot manage his hosts. Each personality has its own way to perceive, to think, and to relate to other people; particularly, their conscience may be totally different, and one personality could be devastated by the actions of the other one. When a person has one of his personalities, he does not always remember what happened when he had in another personality. This disorder has been popularized by the famous novel of Stevenson: *The Strange Case of Dr. Jekyll and Mr. Hyde*.

Having several personalities is not entirely bad: it often happens that humans present different personalities, although they do not have this disease; for instance, a charming friend may harass his wife at home. It can even be good to change personality according to the circumstances: emotion may be pleasant when courting a girl, but disastrous when we are negotiating an agreement with someone who wrongs us. The perfect solution would be to have a personality suited to each situation, and we could change our personality when we seeing how some of our traits may be inappropriate in the present situation. Unfortunately, this is not easy to do: many emotional people have resolved that they would not be angry during an important meeting with hostile participants, but nevertheless they flew into a white rage. Thus, while the possibility of multiple personalities may be useful, the problem is that we are not able to do it well. When it is a disease, this is dangerous because each personality ignores the deeds of the others, which may be certainly embarrassing when we meet someone who met us when we were the "other". Above all, it is appalling to know, as it was the case for Dr. Jekyll, that the other man has committed actions with which we strongly disagree.

Multiple personalities of an artificial being

Artificial beings may have multiple personalities with the same advantages as for the humans, but without the drawbacks that they entail for us. Let us assume that we have written a program that behaves differently according to the value of a parameter P. When P is high, it executes its task very carefully, but very slowly, while when P is low, it will be fast, but its results will have a low quality. When an artificial being uses this program, it can define a more or less high value for P, depending on the importance of the task or the remaining time. Therefore, it will have several personalities suited to the circumstances, in the same way that a husband may be odious with his wife, and charming with his friends. However, it is much easier for a program to juggle between several personalities, it is sufficient to define the value of some parameters for a short while. On the contrary, the natural behavior of a human may appear again because he is led by his emotions.

For instance, as CAIA is a scientist, one of its goals is to create methods for using its rules, so that it will be able to solve problems efficiently. When this task is completed, CAIA has three methods for solving problems: it found two methods and, to help it at the beginning, I gave it a third method. When it solves a problem, its behavior will be very different according to the method that it uses, because they have different characteristics. The first method that it created leads it to execute many rules, even if it is unlikely that they will be successful. Thus, it finds solutions, which may be very elegant: as it has attempted many possibilities, it will rarely miss the application of a useful rule. It will often be able to move toward the solution without backtracking uselessly; however, that may take a lot of time. For its part, the second method, which it had also created, is more selective, it considers fewer rules; thus, it is faster, but sometimes it may miss an elegant solution. Finally, the method that I have given is rather similar to the second one, but not as efficient because I was not able to define the situations where a particular rule will be considered as well as CAIA could. Thus, using each of these methods leads to a different behavior for the solver. With the first of its methods, CAIA is more determined to find an elegant solution, that is more artistic in some way, while it is more efficient with the second method, more concerned with a good use of its time. My method is the most secure because a bug may lead to aberrations in the methods that CAIA generated; so, when there is an anomaly, it is better to see what happens with my method. It is also necessary to regularly compare the results obtained by my method with those of the methods found by CAIA, so that one will be sure that nothing is wrong in the expertise that creates these methods. As they became more and more reliable, the importance of my method has decreased with the years. In all cases, when CAIA solves a particular problem, it chooses its personality.

3.5. The Society of Mind

In his book, *The Society of Mind*, Marvin Minsky explains that the human brain may be modeled as a society of relatively independent agents, where each of them is specialized in the execution of a particular task. In the preceding section, Dr. Jekyll and Mr. Hyde competed for control of their host; here many individuals collaborate so that, together, they succeed in accomplishing the tasks that they have to fulfill.

For its part, during one of its lives, CAIA includes five agents, each of which is specialized and works with the other ones. These agents are the solver, the monitor, the manager, the adviser and Zeus. The solver solves constraint satisfaction or arithmetic problems. The monitor observes the behavior of the solver, and takes into account the difficulties that the solver is unable to deal with. The manager is a scientist: it chooses the tasks that it will execute, particularly it gives the monitor the problems that must be solved. This agent is the core of CAIA, it determines what direction will be given to its research, and it implements it. The adviser observes the behavior of the manager, as the monitor does for the solver. For the present time, its part is rather limited, it regularly assesses what happened during a period of about one day. However, it has very good prospects since, from the bootstrapping point of view, it will develop until it will be able to replace me completely. Finally, Zeus is a safeguard, it supervises the whole operation, and attends to the proper working of the artificial being; particularly, it may get CAIA out of a loop if it is executing the same sequence of instructions ceaselessly. The links between these actors during a life of CAIA are shown in Figure 3.1: an arrow going from actor A toward actor B means that A can observe what B is doing, and also give it orders. Zeus observes the super-actor consisting of the first four actors. I do not include the leader of the bootstrap in this figure: his role is only to start CAIA, to stop it when enough time has passed, and to examine its results. He does not intervene during a life of CAIA, but very often between two lives where he is leading the bootstrap.

It is interesting to consider CAIA as the grouping of these five agents because the consciousness often corresponds to the observation of one agent by another one. For instance, the monitor regularly observes how the solver is progressing while solving a problem. However, some agents, such as the solver and the manager, also observe themselves. Furthermore, the conscience may also be shared by several actors, one of them can execute a task while another one attends to its course so that it is not doing forbidden actions. A line of pine processionary caterpillars forms a society of agents, every caterpillar is built up to follow the preceding caterpillar, which leaves a silk thread. If it happens that the first caterpillar of a line meets the last one, they will move into a circle for hours. This is due to the lack of an agent observing the line, knowing that it is wrong to loop in that way. Even if each agent has its conscience, it is necessary that other agents can watch it because auto-observation has its limits: a disruption may lead not only to poor work, but to a perturbation of the mechanisms that could observe or correct it.

If an artificial being is composed of several specialized agents, many operations associated with consciousness or with conscience are actually actions made by an agent on another agent. Thus, it is no longer necessary to implement the reflexivity, which hugely complicates the realization of the systems where it appears.

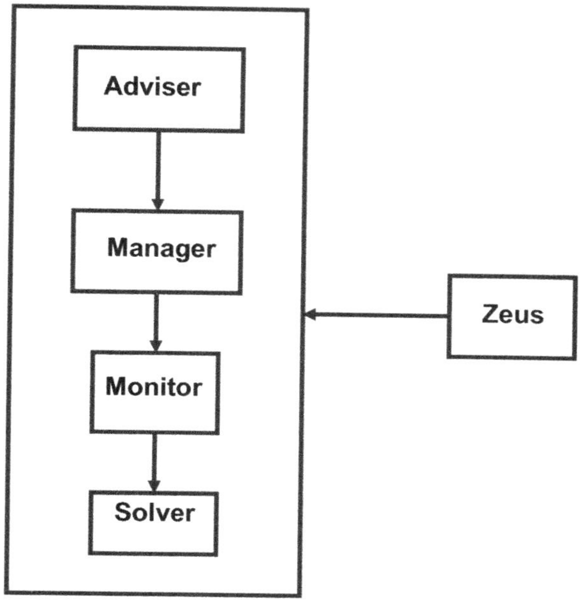

Figure 3.1. *Links between CAIA's agents during one of its lives*

3.6. More on the subject

Jean-Henri Fabre described in Chapter 20 of the sixth volume of his *Souvenirs entomologiques* an experiment where he induces a file of pine processionary caterpillars to circle during a week. Only the cold succeeded in stopping their circular trip.

When I was young, I made the same experiment. It is very easy with a twig to guide the leading caterpillar of the single file toward the end of the last caterpillar; then, they circle for hours. Naturally, one must not touch them, as they have urticating hairs, but these hairs are scattered by the wind, and I had to stop my experiments because I got a terrible rash. Fortunately, computers are not so dangerous, even when they are in an infinite loop.

Chapter 4

Some Aspects of Consciousness

We study consciousness in this chapter and the two following chapters. I begin by describing some of its aspects, then, in Chapter 5, I examine why consciousness can be useful for an artificial being, and finally, in Chapter 6, I will show how it can be implemented.

Consciousness gives us information on the events that occur in our brain; in an analogous way, it gives an artificial being information on its processes that are running, on its present state and on its knowledge. Our ability to know a part of our thoughts improves our performance considerably. First, that gives us more possibilities for learning because we have some idea of why we have made a disastrous decision; thus, we can avoid taking a similar action in a similar situation in the future. In contrast, we can favor the actions that lead to excellent results. Furthermore, as we can access our thoughts almost immediately after they occur, we can immediately see if something is wrong; for instance, if we see that we are going round in circles, or that we do not get the results that we hoped for, we can correct the situation quickly.

The black box is the opposite of consciousness, and we must always avoid it. Many things happen in a black box, but we have absolutely no idea what these processes are; it only produces a decision without any explanation. This is unpleasant, but above all it is dangerous because, if the decision is wrong, it is difficult to examine it critically: we do not know the elements that led to it. Let us assume that a physician tells us that one of our legs will be amputated, while it seems in perfect working order and does not cause pain. Perhaps, he is right, but we want to know why he reached such a decision: the brain of a physician must not be a black box for his patients. Intuition is a mechanism which is a black box, even for its host, and many are reluctant to follow their intuition because they do not know why it has arrived at its result. Unfortunately, we cannot avoid the fact that some parts of our brain will always black boxes for us; in contrast, we will see that artificial beings can examine any of their own processes if they want to.

As many aspects of oneself may be observed, we will begin by examining six of them. Then, we will show that the possibility of self-observation has limits, for humans as well as for artificial beings.

4.1. Six aspects of consciousness

The definition of consciousness includes several kinds of self-observation for human or artificial beings; in this section, we consider six of them. First, one can know that one exists, that one is in an active state, in contrast to deep sleep or a coma for a human. Then, one can know what mental actions were executed during the completion of a task; this is dynamic information, which bears on a succession of changes occurring in our brain or in its processes. Moreover, one can know in what internal state one is, and this information is static. Next, one can also access to the knowledge that one uses to perform a task: it is important to have a model of his/its capacities and inadequacies. Finally, one can feel that one is an individual, different from the other individuals who/which are our relations.

4.1.1. *One is in an active state*

A meaning of the word "conscious" is used to refer to the state of a man who is neither sleeping nor in a coma. At first sight, an artificial being would always be conscious in this meaning when it has been started, and has not stopped because it has completed all its tasks. Actually, there are two exceptions: firstly, the leader of the bootstrap or the operating system might stop it momentarily, and secondly it may be looping without giving the control to one of the processes that can observe its behavior. We will consider these situations where an artificial being is in such an unconscious state.

The computer may be temporarily stopped

After I launch CAIA in a new life, I often need the computer for other tasks, for instance I want to send an email or write a paper. Then, several processes are executed on the same computer and its operating system will allocate time in turn to each process. Thus, each of them, including CAIA, is stopped, then restarted exactly in the state that it was in before it stopped. CAIA will not see that it has been stopped if it does not try to seek it, it is not aware of these interruptions. It is like some epileptics, who are stopped in the middle of an activity by a seizure, and who restart as if nothing had happened; in some cases, they do not even know that they have suffered a seizure. However, CAIA needs to know the time necessary for completing a task, so that it can compare the efficiency of the various methods that it can use. To do that, it can look at the time before and after the task; the computer has an instruction that gives the number of seconds since January 1970. However, the difference does not give the wanted time: if this value is 5 minutes, it is possible that the computer worked for our task for only 2 minutes, thus CAIA was sleeping for 3 minutes. Fortunately, the operating system can exactly measure the time allocated to each task, and not only the chronological time between its beginning and its end. This is essential because, if the measure of the time is erroneous, it will wrongly judge its experiments since its evaluation strongly depends on the time necessary to complete a task. Even when it is the only task running in the computer, there are short sleeping periods, while the operating system is working.

Moreover, I may have to stop the computer for various reasons, and there may be power cuts. Therefore, I cannot be sure that I can have CAIA running for more than a month without any interruption. Sometimes, I have to stop the computer, and I restart CAIA later. To do that, the operating system saves its state; when CAIA restarts, it resumes from this backup. The artificial being may be frozen for as long a time as needed, and restart as if it had never been interrupted.

A process may enter into an infinite loop

CAIA may also lose consciousness when it is looping. While it is in a loop, it is possible that there is in the loop a module where it is observing its behavior; this means that it is regularly regaining consciousness: in this situation, it sees that it is in a loop, and it can stop it. However, a loop usually occurs in a small program, the loop includes few instructions, and none is a step where CAIA observes itself. Thus, it becomes completely unconscious, and this could last for months. Unlike the preceding situation where it was sleeping, the computer is working for CAIA, but this work is completely useless. It no longer knows what it is doing, and it could not even take the decision to stop this task if it did not have an agent, called Zeus, which can deal with this kind of difficulty. Zeus is regularly activated by the computer clock because CAIA has asked the operating system to interrupt it, for instance after five minutes, and then to give the control to Zeus. Thus, five minutes later, Zeus will be automatically started, whatever happens with CAIA, even if it is in an infinite loop. Then Zeus examines what has been done since the last interruption: if it has begun a new task, or if the preceding task has progressed, it is certainly not in a loop. In contrast, if it has remained in the same state, if it has applied no rule for solving the current problem, it is more than likely that it is in a loop. In this situation, Zeus decides to stop the task, after storing exactly where the loop occurred so that CAIA may later understand what happened. Then, it restarts CAIA, which in this way regains consciousness; it will find a new task, but it knows that it was trapped in a loop, the task where it occurred, and even at what step in the process of this task the problem occurred.

This ability to regain consciousness after losing it is essential, as without it the unconscious stage could last for ever. This is particularly important for an artificial being, which is not linked to the exterior world. It also happens that animals can begin a loop; if we show a fly the same pattern 1,000 times, it will give the same answer 1,000 times. However, these loops may stop because tiredness or a change in the world will lead it to give another answer, and thus to leave the loop. We have seen the example of the pine processionary caterpillar; they do not turn in a circle forever because, after some hours, a perturbation will always break the loop. In contrast, in the virtual environment of CAIA, these modifications of the environment never occur. Thus, a loop could last endlessly if the clock did not give control to Zeus, which ensures that CAIA will have a conscious period whatever happens. We can notice that other aspects of the consciousness play a part in this process: Zeus knows what CAIA has done since the last interruption so that it can find whether there has been a progression. It has also to know where the loops happened, so it can store the characteristics of the task, and of the situation where the loop occurred, particularly the module where it was looping. In this way, CAIA will not repeat the same mistake, and later on it will be able to understand the reasons for this event. The auto-repair is limited for the present time to gathering more information on the events that lead to this loop: it executes the task again, but this time it creates a detailed trace of its steps. With this trace, the leader of the bootstrap will easily find the reasons for this loop and finish the repair. If, unfortunately, CAIA is clumsy enough to re-enter the same loop, Zeus will still be able to break the loop; anyway, more information will have been gathered on the causes of the loop. This capacity for escaping from a loss of consciousness would be very useful for humans: without exterior help, we could wake up from a blackout or from a coma.

A partially conscious state

Curiously enough, CAIA has a state other than being conscious or completely unconscious: for a human being, unconsciousness entails the impossibility of having access to its memory or to fulfill any task. In contrast, in this third kind of state, partial consciousness, CAIA has access to its memory, it can give helpful answers to questions, but it is no longer conscious of what it is doing. This happens in the following circumstances: when I consider that it has run for long enough time, I stop its research activity forever because it is more useful to modify it, take into account its results, and to restart a new variant of CAIA. However, in order to write this book, for instance, I may need more details on what happened during its life, such as the list of the problems where a clever method was significantly faster than a classical combinatorial method. Then, I awaken the older version of CAIA, without allowing it to resume its research; I only need some information generated from the facts that it kept in memory. Some of its modules are therefore run again, particularly those that analyze its results so that it can answer my inquiries. It can perfectly fulfill some tasks, but others are no longer allowed. It is thus in the same condition as someone who has suffered from a stroke: it can carry out some tasks perfectly, but it is completely unable to perform some other ones. In this situation, CAIA is active and performs useful activities efficiently, although it is unconscious; it can observe itself only when it is solving problems or managing its life as a scientist.

To summarize, an artificial being may become unconscious in several ways. First, CAIA has the life of an epileptic, but this does not hamper it because the interruptions coming from the operating system or from the leader of the bootstrap do not matter for a virtual being as long as one measures the effective time needed to fulfill a task. However, for a robot interacting with the real world, such interruptions would be unacceptable if they lasted too long; fortunately, it is possible to indicate to the operating system that the duration of an interruption has to remain negligible. In contrast, entering an infinite loop may have serious consequences, and an agent, Zeus, has been specially designed for righting this difficulty. As one cannot ever be sure that a program will never loop forever, it is essential to build artificial beings so that they can recover by themselves from such loops.

4.1.2. *One knows what one is doing*

It is useful to know what is happening in our brain, or what steps have been executed by the computer: in that way, one can repeat it, teach it, correct it, or avoid doing it again, depending on whether the result was right or wrong. In this section, we will particularly focus on human behavior, to show the interest and the limits of this possibility. As we are used to these limitations, they seem natural to us, we underestimate their importance, and we are not aware that artificial beings are not restricted by most of them. This ability to observe one's own thinking is as crucial for artificial beings as for human ones.

We know the physical steps that we have carried out to modify the exterior world, for instance, how we have cooked a cake. However, we also know some of the intellectual steps that were carried out in our brain when we planned and executed this recipe. Instead of remembering that we have broken an egg, we remember that we have chosen to include only one egg, rather than two as was indicated in the recipe, because our cholesterol level is too high. Many events occur in our brain before some of them lead to more visible actions on the world. In that way, we have a dynamic vision of ourselves, we know a sequence of physical and intellectual actions that we executed while performing a task.

It is important that we can be conscious of some of the steps which we are executing to fulfill a task; this enables us to teach our methods to other people so that they rapidly become successful, and to check what we are doing. If we find that an intermediate result, which is the value of a probability, is greater than 1, we can be sure that we have made a mistake somewhere, and we stop so that we can investigate where it happened. To find this mistake, we also need to know other intermediate results. Thus, only obtaining the result would have been inadequate: it might have seemed plausible, and even if it was not, we would be unable to find the cause of the mistake if we could not access to some intermediate results. Usually, in mathematics, we know many intermediate results but, unfortunately, we know none for the results given by our intuition.

Naturally, as Marvin Minsky has mentioned, consciousness does not occur simultaneously with the processes that we observe, but slightly after. When we use it, we are observing the past, but a very recent past. In the same way, when an artificial being observes its own processes, it does not do it simultaneously, but with a small delay.

An experiment with chess players

It is interesting to examine what we know and what we ignore of our own mental processes. Let us consider a chess player who is choosing his next move. He knows a large part of the elements that he considered while he was thinking. In the following example, the reader who is not a chess player can understand the kind of information which is considered by the expert; however, for chess players, I also display the chess board so that they can understand more completely how a player analyzes a position.

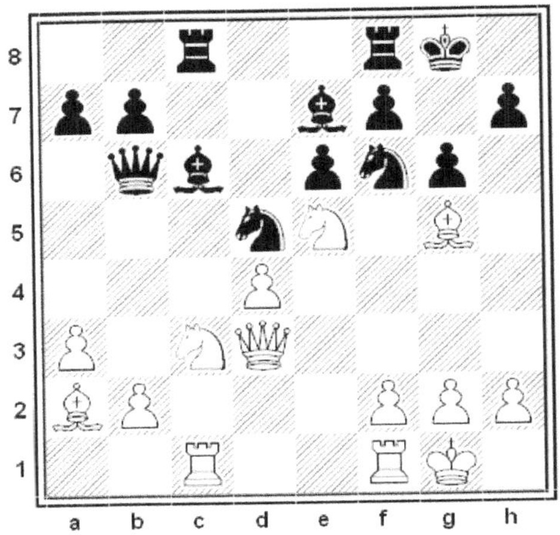

Figure 4.1. *White to play. De Groot asked several chess players to think aloud while they were choosing the best move in this position*

When Alekhine, then the world champion, received the position given in Figure 4.1, he mentioned:

– some moves which he has considered:

"N–N4 comes into consideration."

– why he has rejected some of them:

"NxN maybe; 1...NXN; 2. BXN, BXB/N with many capturing possibilities, but on first sight not convincing."

In the algebraic notation, these moves are: 1. Nc3xd5 Nf6xd5 2. Ba2xd5 Bg5xe7.

– some characteristics of the position, which he has evaluated:

"Is the Pawn at QN2 really attacked?"

– some similar situations which he remembered:

"At first sight there is a dark memory of a tournament game by Botvinnik–Vidmar (Nottingham)."

The preceding quotations are given in Adriaan de Groot's book; they appeared in the protocol by Alexander Alekhine, who had to think aloud while he analyzed this chess position. In the present case, the instructions were to say all of the ideas the subject was conscious of while he was determining the move that he would play in the given position. Naturally, thinking aloud cannot describe the unconscious processes, and the obligation to speak may disturb the thoughts of the subject. However, it seems that some subjects succeed in giving most of the ideas they are conscious of when they are used to this constraint.

This protocol is interesting because it gives an idea of the processes running through the head of a world champion, while he is thinking about the move that he will play. However, we can notice that much information is missing:

– Why did he contemplate the move N–N4 (Ne5–g4)?

– Why did he think to a position of a match Botvinnik–Vidmar?

– Why did he especially consider the Pawn on QN2 (b2)?

Many other characteristics of the position are not in this protocol. Perhaps Alekhine did not consider them, or had not the time to verbalize most of his thoughts, or they were unconscious. Even though we can access some of our thoughts, it is difficult to memorize all of them. This is why thinking aloud is useful; it allows us to create a trace of our thoughts, to be studied later. It seems that there are large differences between subjects; some produce a very rich protocol; one reason may be that thinking aloud is easier for subjects whose thinking uses a natural language. In the experiments made by de Groot, it is particularly intriguing to compare the protocols by Alexander Alekhine with those by Max Euwe; they twice exchanged the title of world champion. Euwe's protocol is much richer. After some general considerations on the position, he says: "Let us look at the consequences of some specific moves." Then, he successfully considers four capture moves. For the fourth one, BxN/5 (Ba2xd5), he develops a sequence of eight moves, and then there is a surprise: "But that's good for white." Finally, he decides to play it. In contrast, Alekhine had immediately contemplated this move, and he had decided, before considering any other move: "1.BxN/5 is the one enticing move." Then, he had only made a quick verification, considering two other possible moves. Both players decided to play this move, and later, after a thorough analysis, everybody agreed that it was the best choice. We see a huge gap in Alekhine's protocol: why did he immediately consider the best move? Euwe's protocol is clearer: he systematically developed what may happen after all of the capture moves, and then he deduced that one of these moves was very good.

We are lucky to have the possibility to access a part of what happens in our brain but, unfortunately, we cannot access a large part which involves important processes. In Particular, de Groot's experiments established that the strength of the best chess players come, for a large part, from their ability to consider the best moves. For one particular chess position, the five grandmasters who were his subjects had considered the best move, while many good players had not played it because they had not even considered it. Although we have a protocol of Alekhine's thoughts, the mechanisms which enable him to consider the best move are still mysterious, we only know why he has judged that it was a very good move. This gap is very serious for us; AI scientists cannot know what mechanisms are used by the best players when they are choosing the moves that they consider; so, we cannot implement them in our chess playing programs, although they are essential for the performance of the human players.

Many of our mechanisms are unconscious

In domains other than chess, it is also very difficult to know why a human has made a particular choice. Statistical studies have shown that the reasons given by the subjects are actually false. In an experiment, four nightgowns were presented to women, with different presentation orders, for instance one particular nightgown was at the right one time out of four, in the second position also one time out of four, etc. A statistical study has shown that the subjects had a strong preference for the nightgown at the right. However, the subjects flatly denied that the position of a nightgown could have the least influence on their choice. This is normal, as we have a very imperfect knowledge of the mechanisms that are running when we are making a choice: the subjects knew that they liked the color of a particular nightgown, while they had not liked the cut of another one. However, they had no way of knowing how they were influenced by their position in the presentation.

The limitations of our consciousness are specially evident for perceptual mechanisms, which are very fast. What criteria are we using to recognize that a particular character is the letter "A"? We can give some characteristics of this letter, but that has nothing to do with the method that we are using to identify it. There will always be prints of this letter that we correctly identify as an "A", although they do not satisfy all the characteristics that we have defined. Because we are unable to indicate how we are recognizing handwritten characters, this task is still difficult for programs. This is why some web sites ask us to recognize badly written text (called captchas) when they do not want automatic systems to log in. Captcha is an acronym for **C**ompletely **A**utomated **P**ublic **T**uring test to tell **C**omputers and **H**umans **A**part; it is a test that humans can easily pass while computers still have very poor performance at this.

The following example shows how it is difficult to know our own perceptual processes. Many years ago, an article in *Scientific American* presented two photos, apparently identical, of a young girl, and the author asked the reader to find the photo where the young girl was the most beautiful. This question seemed stupid, since both looked exactly the same, however it was obvious that she seemed much more beautiful in one picture. The explanation was that this photo had been retouched so that the pupil of her eyes had been dilated. Men prefer that women look at them with admiring eyes, that is why Venetian women dilated their pupils with belladonna, whose name comes from the Italian words "bella donna", meaning "beautiful woman". All the male readers unconsciously perceived the difference, since they used this information for their choice; however, they were unconscious that this difference existed.

We cannot be as conscious of our intellectual mechanisms as we would wish

There are many causes for the limitations of our consciousness. Firstly, we are disturbing ourselves when we are observing us while we are acting, thus we do not behave in the same way as we would do if we were not observing ourselves. This is the problem of the bearded man who could not sleep because somebody asked him whether his beard was above or under the sheet when he was sleeping. Moreover, we are memorizing what happens while we are acting, thus we have to do three things simultaneously: to act, to observe oneself, and to memorize these observations. Our brain is already very inefficient when we have to perform two tasks simultaneously, three is more than we can do. Thinking aloud avoids the necessity of memorizing, but it disturbs us more or less according to the subject, and we can enunciate only the verbal part of our thinking, which may be minor for some of us.

However, the main limitation comes from the fact that most of our mental processes are unconscious, that consciousness will never access. What we call "intuition" is only a completely unconscious process. We can see here one more reason why consciousness is so useful: it is difficult to act from an intuition because we do not know the steps that justify this intuition. If my intuition tells me that a particular candidate to a job will not do, can I follow this advice to disrupt his life by rejecting his application, although his qualifications are perfect? Perhaps, my intuition is only based on a physical likeness to someone who was a serious nuisance. When we do not know the reasons for a decision, it is natural to mistrust it, it can be based on facts as irrelevant as the position of a nightgown at the right. Moreover, it is difficult to have such a decision accepted by other people, because we have no arguments to justify it. However, it happens that our intuition produces some surprisingly good results!

Blindsight also shows the importance of the unconscious processes in our brain. In that situation, some blind people consciously have the impression that they see nothing at all; however, their answers to the questions on the images that are presented are not completely random. For instance, they often can give the position or the kind of motion of a visual stimulus when they have to guess it, and they wrongly believe that they are giving random answers. Some parts of their brain receive the information, although they do not know it, but this contributes to a correct answer without any intervention by conscious processes.

We may forget to use our consciousness

Another limitation of our possibilities for observing ourselves is that we do not use it often enough, even when it is possible. While we are thinking, we ought to make an effort to examine more often what we are doing and memorize it, but we are too much taken up by our thoughts. For instance, in order that a chess game does not last too long, each player has a clock, which indicates how much time he still has before playing the fortieth move of the game. If he is wasting too much time on the first moves, he may have only two minutes for playing the last ten moves: against a strong player, he would be in a catastrophic situation. In some games, players think for more than an hour on a single move. If they had spent only 45 minutes, the quality of this move would not have been lowered too much, but that would have given them 17 minutes for the last ten moves, and the quality of their game would have been considerably improved. Yet, some players are so deep in thought that they do not take the time to examine their thoughts, although that would enable them to stop thinking about this move, and to keep enough time for the last moves. Monitoring is just this action of constantly considering the current mental process and judging it; this allows one to change the subject of the thoughts, or to decide that it is time to take a decision. We do not do it easily because we do not have an efficient mechanism for interrupting our thoughts so that we could observe them.

Artificial beings may have an adjustable consciousness

We will see that artificial beings are not as restricted as ourselves when they are observing themselves. It is never good to observe everything, one would lose a lot of time on events that do not matter. Fortunately, an artificial being may choose to observe what it wants among its own processes, without any limitation. It may have an adjustable consciousness: it can choose to observe one of its modules, and it can do it in a way as detailed as it is useful. Then, when it has observed what it wanted to, it can momentarily stop observing until it launches another module which it considers as important to watch. We will later show that two techniques allow it to obtain this result; they were developed by computer users, who wanted to know what happened in their programs, so that they could debug them. Here, the only difference is that an artificial being, instead of a programmer, analyzes the observations gathered while using these techniques. Moreover, an artificial being can easily memorize what it is observing, it is sufficient to put it in a file, which will contain a trace of the interesting events. It will analyze this trace later, but it can also examine any event immediately, and instantly take some actions if needed, thus monitoring the current process. Storing these data do not at all perturb the process that observed them. Considering the importance of this kind of consciousness for humans, artificial beings potentially have a huge advantage. Unfortunately, we still have to find how to use this advantage; this is difficult because our only model is human consciousness, and we have no example of humans having the extended capacities of artificial beings.

More on the subject

Adriaan de Groot wrote the fundamental book on the mental processes of chess players: *Thought and Choice in Chess*; this text stems from his 1946 thesis. His subjects were players of various strengths, from skilled players to world champions. In particular, Max Euwe, then the former world champion, agreed to participate in many experiments. Using the thinking aloud method, the subjects had to say what they were thinking while they were choosing the move that they would play in a given position, or when they were recalling a position that they had seen for only a few seconds. This work was one of the first using the thinking aloud method, and it is still noteworthy for the exceptional quality of some of its subjects.

4.1.3. *One examines his/its internal state*

At any time, an artificial being can stop and completely examine its internal state at that particular instant. For instance, it will see that there are the following succession of current processes, each one being a step toward the completion on the process above it:

– It is solving the crypto-addition DONALD.
– It has just generated the new constraint 10R+T+100 = 2D+10L.
– It is checking whether it had not already found and stored this constraint.
– It is verifying that it is different from the already stored constraint 10G+8D+T = 20L.
– It is comparing the terms 10R and 10G.
– It is checking whether the variables R and G are the same.

Naturally, it will see that R and D are different, thus both constraints are different, and it will compare the new constraint with another stored constraint, but this will happen later. Here, we are only interested in the description of the current state of CAIA.

All the processes currently working on the present task are frozen so that it can observe all of their characteristics, store them if it wants to, and then restart them as if they had never been stopped. This static aspect of consciousness is different from the dynamic aspect, which we have considered in the preceding section, where we were interested in the sequence of actions that it performed.

We have difficulty observing the state of our brain

We rarely think about this aspect of consciousness because humans have almost no possibility to observe themselves in this way: we cannot stop our thoughts, examine the present state of our brain, and then restart our original line of thought, as if it had never been interrupted. However, this capacity is crucial, the dynamic and static aspects of our thought are complementary. In an analogous way, in a game, a dynamic aspect appears when one considers the succession of changes brought about by the moves, while a static aspect takes place in the examination of the board and of the various relations between the pieces. It is likely that, as for games, the static study of oneself may provide useful information for understanding what happens while one is carrying out a task.

Proprioception gives us a limited knowledge of our current state, such as the relative position of our limbs and the tension of our muscles. Even when we close our eyes, we can indicate the angle between our upper arm and our forearm. However, it is mainly unconscious, as most of the signals sent by our articulations and muscles are used in the cerebellum and do not go higher in the brain. This enables us to control our posture, but is mainly not in the domain of consciousness. The situation is worse for our mental state, we have absolutely no idea of the state of the various parts of our brain. When we are speaking, we do not know that our Broca's area is active. Many centuries were necessary before scientists agreed that our thoughts are generated in our brain: it was only in the 18th century that this was definitively established. A genius as outstanding as Aristotle believed that the heart was the seat of consciousness, and that the role of the brain was to cool down the blood! Hippocrates correctly identified the role of the brain, but the authority of Aristotle prevented, long after his death, the idea becoming widely accepted.

For many neurological diseases, physicians have difficulty knowing what their exact causes are because the patients cannot indicate them. An epileptic does not know the origin of his seizures, and the surgeon cannot act with precision to cure them. It has been noticed that the seizures disappear when most of the connections between both hemispheres are cut; however, a precise localization of their origin would perhaps allow a less drastic surgical cure. New techniques for observing the human brain have led to significant progress, but all these techniques, such as positron emission tomography, magnetic resonance imaging and computed axial tomography, are due to the realization of new devices. We are always unable to observe directly what is occurring in our own brain. Incidentally, even if we had this knowledge, it would not be very useful because we have limited possibilities to act on our brain. For example, if we are not correctly understanding a particular natural language, we could not modify the structure of our neurones, even if we knew which areas are bringing about such malfunctions.

An artificial being can easily observe its present state

An artificial being, such as CAIA, may have completely different possibilities because it can write its modules in such a way that, each time one module becomes active, this fact is stored with the value of its arguments. Moreover, when it executes a module, it can access the values of the variables which are defined or modified. Therefore, if CAIA decides to examine its present state, it examines the stack of the active modules, and it has a complete description of their situation, exactly at the instant where it was stopped. This is as if we could stop our thoughts, see all the active neurones at this time with a complete description of their state, and then restart in our old line of thought, which would not be disturbed. In addition, it can store all of the interesting facts, and even modify some parameters so that the following treatments will be improved. It can examine its state as often as necessary, the only drawback being that too many stops will slow down the thinking. For CAIA, the only difficulty is to know when it will stop, and what it will observe; this decision may be taken by itself or by the leader of the bootstrap, which/who may want to deepen its/his understanding of the use of a particular module. The decision may also be the consequence of a serious mistake pointed out by the operating system. We return to this point in Chapter 6.

Naturally, one cannot observe one's own state too often. CAIA could spend all its time observing itself and analyzing these observations, instead of completing more useful tasks; so it should be done sparingly. At present, this possibility is mainly used for debugging CAIA, by the leader of the bootstrap or by CAIA itself. For instance, when CAIA has seen that an anomaly occurred during the execution of a particular module, it will execute the task again until it comes to this module. Then, it switches into a conscious mode, where it can access all of its own characteristics. If the correct value of a variable is known, it can check whether this variable actually has this value; if it is not true, it has the reason for the dysfunction.

It can also store the values of some interesting variables, and the leader of the bootstrap will examine this file later on. At the present time, when the operating system signals a serious anomaly, it writes into a file the values of all the variables of all the active processes. This information gives a complete description of its present state, and it is very helpful for finding why this incident occurred.

Usefulness of observing its present state

Thus, artificial beings can have a perfect knowledge of the present state of all their active processes, and restart without disturbing their following thoughts. I believe that this capacity gives them an extraordinary potential to achieve tasks that we will never be able to because we lack this possibility. However, as no living beings can observe the state of own brain while it is working, we have no model to inspire us, and it is difficult to see all the improvements that could arise from this possibility. We have three problems to solve before implementing an artificial being that would use the advantages of this approach fully: it must determine when it is useful to observe its own state, choose what it will observe, and analyze these results so that it modifies itself for improving its performance. Let us assume that we could have the same capabilities as these artificial beings, and that, unfortunately, we suffer from an embarrassing disorder, for instance we stutter. If a stuttering patient could observe the condition of the area producing the spoken language, he would follow step by step the state of this area when he is wrongly repeating a syllable. Thus, he could see why this repetition occurred, he would modify the elements that brought about this mistake, and so he would be definitively cured. Naturally, men will never be able to do this, but artificial beings can perform such observations and modifications on themselves. When they are debugging their program, the programmers constantly use this possibility to see the values of the variables at a given moment. Realizing artificial beings that can do the same thing on their own programs is just a natural extension of these methods: the only difference is that the artificial being, instead of the programmer, chooses when it stops and what it observes.

This possibility is not only useful for finding mistakes, it also improves learning capacity. Learning has two sides, observing and modifying; the first part is related to consciousness, and we have just seen that an artificial being can observe themselves in a way impossible for a human. It can, better than ourselves, observe its state, so it is easier to find what goes wrong. Unfortunately, the second part, that is, finding how to modify itself, has still to be developed, but their present limitations for learning do not come from limitations of their consciousness.

An artificial being is not restricted by our impossibility to observe the state of our brain, it can completely observe the present condition of all of its modules. For the present time, this huge advantage is only marginally used because it is not sufficient to know why something goes wrong. It is also necessary to know how to use this knowledge so that an artificial being could improve itself. The static observation of a system by itself is not a science fiction story, it is widely used for debugging, and CAIA uses it. At the moment, it uses this capacity only to help its leader of the bootstrap to debug it. The learning depends on the couple (self-observation, self-modification), and only the first part has been implemented so far. However, as its capacities for observing its state are greater than ours, we hope to have very good performance when the second part is implemented.

4.1.4. *One knows what one knows*

The basic idea of expert systems is propitious: they try to use the fact that human experts are very successful in their domain. Therefore, if we ask them their expertise, and give a computer system this knowledge, this expert system will be as successful as the human experts. Unfortunately, after many attempts, it seems that is not easy to implement this appealing idea; one of the reasons for this difficulty is that a human expert cannot express most of its knowledge.

An expert does not always know what he knows

The restriction on the knowledge of our own knowledge has considerably hampered the development of expert systems. Only a beginner can say what they know, because they have just acquired it, and it is still in a declarative form. In contrast, human experts have proceduralized their knowledge, so that it can be used efficiently. When we describe to them a situation where there is a problem, they perfectly know what has to be done, but it is working on the problem that triggers the access to the useful knowledge. Most of their knowledge is in a procedural form, it is not stored in a file which they consult when they need it. For an expert, knowledge is ready to use, hidden among other bits of knowledge; it will be automatically triggered by some characteristics of the current situation. Thus, an expert is unable to enumerate his own knowledge when we ask him to do so. He will invent rules that seem plausible, but they are not those that he really uses because he is not conscious of this knowledge: we mainly have access to our declarative knowledge. For instance, we know that the letter "A" is the first of the alphabet, we can describe it by two straight lines, inclined at 60°, which met at their top and a horizontal line linked them in their middle. However, this does not give us the expertise that we use to recognize this letter, we are able to do so even when the preceding definition is not satisfied: the lines may have interruptions, the horizontal line is inclined, it may cross the other lines, etc. For every description that we will give for this letter, we will find examples, which we recognize as the letter "A", although they do not agree with our definition. Actually, we have invented this definition from some examples that we identify as letter "A". We have no access to the procedural knowledge in our brain, which enables us to perceive the pictures, including the letters. However, we are experts in letter recognition, as soon as we can read easily. This situation arises for all human experts, in medicine as well as in mathematics, architecture, chemistry, and so on.

When an expert carries out a task, he uses procedural knowledge that he has learned directly in this form, and there is no declarative version which he could express. We have a relatively good access to the events of our life, but a poor one to the directions for use of our knowledge. At the start, a beginner receives a partial set of declarative knowledge, such as the mathematical rules that we have just seen. He finds a first method for using them; however, in the future, he will usually directly modify his procedural knowledge rather than adding new bits of declarative knowledge, and translate again the whole of it. This is why much of the knowledge does not exist in a declarative form, and why both kinds of knowledge, when they simultaneously exist, are no longer linked. Thus, we are frequently using knowledge which we do not know; if we could examine it, we would probably see that a part of it is defective or obsolete. Moreover, even when it is excellent, we cannot transfer it to beginners, so they have to observe how the experts are working. This is a rather lengthy and inefficient way to learn because it is difficult to understand why an expert chooses to perform a particular action.

It is easy to analyze declarative knowledge

Computer programs, the description of an itinerary, or a cooking recipe are procedural knowledge, which is useful for carrying out a task, but it is not a form easy to analyze. It is practically impossible to understand a program written by another programmer if it does not include declarative knowledge expressed as comments. It is not enough to have access to a bit of knowledge, it has to be stated in such a form that one can study it; this is not the case for procedural knowledge. We have already seen that it is more convenient to have knowledge in a declarative form; then, a translator will translate it into a procedural form, so that it can be used efficiently. Both forms are accessible to an artificial being, but it uses the procedural form for fulfilling a task, and the declarative one for analyzing or modifying its knowledge. In this way, CAIA can be conscious of everything it knows, without any exception since it generates all its programs from its own declarative knowledge: there is no instruction in the programs constituting it which has not been written by itself.

An artificial being can easily enumerate its declarative knowledge, since it can access all the files that contain it. Nevertheless, the most important fact is that it can analyze it, so that it can determine how it will use it: it can proceduralize its knowledge skillfully. For instance, when it receives a new mathematical formula, this capacity enables it to find in what way it will use it. Let us take for example the following mathematical formulae: A*0=0 and A+0=A. In this way, they are given in a declarative form, we do not indicate how to use them; the first one only says that, if somewhere we find an expression multiplied by 0, we can replace the whole by 0, and conversely, if we find 0, we can replace it by any expression multiplied by 0. The formula does not indicate when it is interesting to apply it, using one of the preceding ways, and when it is better not to apply it at all. A clever system can define some directions for use when it analyzes these rules. A general principle states that it is normally useful to reduce the size of mathematical expressions as much as possible. Thus, the system may decide to replace each product including a zero by zero (from the first rule) and to remove every occurrence of zero in a sum (from the second rule). It still has to determine when these new procedural rules will be executed; naturally, it will decide to include them in the procedure that simplifies mathematical expressions. This procedure transforms $X^2+X*0=7$ into $X^2+0=7$, and finally into $X^2=7$. In contrast, it is rarely useful to complicate a formula, thus the system decides that it will never use the other possibility: what is the interest in transforming X^2-2*X+1=0 into X^2-2*X+1=X^7*0? In that way, one proceduralizes knowledge that was initially declarative. Very often, this proceduralization step is unconscious for a human, thus there is no remaining link between a procedural rule and the original declarative knowledge that was used to build it. In contrast, the procedural knowledge of an artificial being can be linked to its original declarative knowledge.

Therefore, an artificial being can express its procedural knowledge, and it can also express the declarative knowledge that was used to generate the procedural one. We have already considered this point when we have examined how it is easy to multiply artificial beings. CAIA executes the three steps of the following plan: gather more declarative knowledge, translate it into a procedural form, and use it. It resumes the translation each time that it receives new bits of declarative knowledge. Therefore, it may have a complete access to its declarative knowledge as well as its procedural one.

A human expert knows many useful facts and methods but, as he mainly stores procedures in his brain, he can neither examine nor share them. In contrast, the knowledge of an artificial being may be acquired or found in a declarative form so that it can easily analyze it, modify it, proceduralize it, and provide it to other human or artificial beings.

More on the subject

Wikipedia gives an excellent introduction to expert systems:
http://en.wikipedia.org/wiki/Expert_system

4.1.5. *One has a model of oneself*

Human beings have a model of themselves, they know their strengths, their weaknesses, and their preferences, they have an idea of their aptitude to solve problems. This model is often inaccurate, but it always exists. For instance, someone knows that he has a tendency to be angry, that he is a good solver of mathematical problems, that he is hopeless at writing letters, and so on. This knowledge on our aptitudes enables us to foresee what is likely to happen when we execute an action. If we foresee undesirable consequences, we will change our plan: someone suffering from vertigo will not try to climb Mount Everest. We can also try to reduce our defects: Demosthenes stammered, so he trained himself to speak with pebbles in his mouth. In the same way, some timid people force themselves to speak to one unknown person every day. A model is not always accurate, so one may be surprised when an action has unexpected consequences; the surprise comes from a difference between the predictions of the model and what really occurs. For a scientist, surprise is a crucial event, it shows that the model has to be modified, thus it can use these circumstances to improve its model.

For living beings other than man, this model of their abilities is implicit: an animal does not know how many meters it can jump but, in front of a ditch, it perfectly knows whether it can jump over it. Even with humans, the model may be implicit: in every situation, a driver knows whether he could stop if there were an obstacle, even if it does not know his stopping distance. However, our model is partially explicit; we know if we are lazy, and if we can write C programs.

A robot which explores space may have faults during the several months, or even years, of its mission. It therefore has to be able, firstly, to see that there is a fault, and secondly to correct it. These problems are well known in AI, and can be solved when the robot can use a declarative representation of itself. Thus, it can see that there is a fault when its behavior disagrees with the behavior predicted by its model. To explain this fault, it examines what it would do when its representation is modified. When a modification leads to the same behavior as the observed one, the fault is explained. In that case, it is sufficient to remove this accidental modification. The problem is easier when a spacecraft is navigating though space, because the environment is not changing too much. In contrast, on a planet, it is difficult to find whether an anomaly comes from a faulty component, or from the distinctive characteristics of the environment.

Predictions, hopes and musts

CAIA does not have an explicit model of its qualities: it does not know that it is obedient and not lazy. However, it uses an implicit model of what normally occurs when it solves a problem; this allows it to make various projections on what will come about in the future steps. These projections may be predictions, hopes and musts.

A prediction can be made while CAIA is solving a problem. For instance, it noticed that it has developed 80% of the search tree, and that it has already spent 75 seconds. In that situation, it will estimate that the resolution will be most likely completed in 35 seconds, for it is taking a safety margin. Besides, it can make other predictions at the beginning of a resolution: let us assume that it has already solved in 200 seconds a problem with the clever method that it has created. Then, it will predict that at most 200 seconds will be necessary to solve the same problem with a combinatorial program, which it will write. This prediction is made because usually the solutions obtained by a clever method are more elegant, but require more computer time. If a prediction comes true, CAIA does nothing, it is normal that predictions are fulfilled. In contrast, if the prediction proves to be false, it notes that event in a file. This is a surprise, it is important to note it but, for the present time, CAIA still leaves it to the leader of the bootstrap to interpret it.

A hope indicates that one will be very satisfied if a desirable but unlikely event will take place. CAIA will hope that a difficult problem will be solved in less than 55 seconds or that the tree generated for solving it will have less than 27 leaves. These hopes are created from the experiments that it has already made with a problem, where it has already tried several methods to solve it. As an athlete always hopes to break his own record in jumping farther, CAIA hopes to improve the way it is solving a problem, and specially to solve a previously unsolved problem. Processing a hope is the contrary of processing a prediction: if a hope is not satisfied, this is normal, it does nothing. However, if a hope is satisfied, this is an excellent surprise, and it is noted in the journal where it keeps all the important events: it did not dare to believe that it could happen but, naturally, it noticed it when it occurred. The assessment made by the adviser takes into account all the satisfied hopes: the more hopes were fulfilled, the more it is happy with the performances of CAIA. When I am writing "happy", this only means that it adds a good mark to its appraisal of the performances of CAIA. For each period that the adviser agent considers, usually a day, these marks enable it to make a global evaluation, which it justifies by the list of the remarkable events, which naturally include the fulfilled hopes.

A must states that a particular event must happen: the solver must have developed new leaves in the tree in the 120 next seconds, or the number of solutions of a problem must be equal to 7. This last must is made when the problem has already been solved by another method, and 7 solutions were found; the number of solutions for a problem must always be the same, whatever method is chosen. The normal situation for a must is that it becomes true: when this happens, nothing is noticed. In contrast, when a must is not satisfied, this is noticed in the journal including the important events. Sometimes, CAIA can analyze the situation, for instance when the number of solutions changes when a problem is solved with two different methods. If so, it determines which resolution is erroneous. Either one has missed a solution, or the other has an incorrect solution. However, it is usually the leader of the bootstrap who has to understand while a must has not been satisfied.

Predictions and hopes are also used when CAIA chooses the next task that it will carry out: it favors those where there is a prediction or a hope. It is better to give a priority to the tasks which can surprise it. The predictions, the hopes, and the musts can lead to an unexpected event, which is a good surprise if it is a satisfied hope, and a bad one if it is a prediction or a must that is not satisfied. Deepening the reasons for these surprises is a difficult task, and CAIA usually cannot do it yet. However, the possibility of being surprised is already useful, these events are pointed out to the leader of the bootstrap: this is an important part of the collaboration between the artificial scientist and the human scientist. The ability to be surprised depends on the existence of a model of itself, which describes what will normally happen.

4.1.6. *One knows that one is different from the other individuals*

Everybody is conscious of being an individual, different from the other individuals that we meet. We are creating models of the other humans as well as of ourselves. This is useful for foreseeing their actions, so that we can avoid interactions with other people that might lead us to unwanted situations.

CAIA is completely lacking in such capacities because it interacts with only one person, its leader of the bootstrap; moreover, it does not even know that this leader exists as an individual! It cannot ask him questions, all the dialogues are between its own agents: the solver can make a request to the agent that monitors the progression toward the solution. Then, the monitoring agent can ask for advice from the manager, which asked it to solve this problem. The links between these agents are stereotyped: for instance the adviser regularly examines what occurred since its last intervention, and informs the manager of its assessment. Most of the multi-agent systems have implemented much richer links between the agents, particularly when all these agents have similar capacities. If so, they can negotiate, vote, bid for a contract, so that they can define together which agent will execute a new task, depending on the burden of their present work and on its abilities.

I created CAIA, I launched each of its lives, but I appear nowhere as long as it is living; it is acting on its own decisions, I will again play a part only when I decide to stop it. This task could even be given to the adviser agent, which might stop everything when it sees that there have been too few interesting events in recent days. CAIA is the only individual operating in its artificial world, it describes its activities into a file because it has been ordered to do so. It does it in the same way that we are programmed to shout when we are hurt, although this is useless when we are alone. Everything goes as if it was on a desert island for its entire life, thus it cannot feel different from another being. It does not even try to model its leader of the bootstrap, since it does not even know that I exist.

Although CAIA has already completed several lives, it does not know its past: the individual that it was in these preceding lives does not exist for it. Its only memories of its previous incarnations are implicit: for example, it knows that some problems are difficult because it graded them so before. Besides, at the beginning of a new life, when it had not yet created a method for solving problems, it temporarily uses a method built in its last life. However, it does not know that its knowledge has two origins: myself and itself in a previous life. Perhaps, in improving its capacities, it could see that something is above it, that there are things that it does not control. For instance, if it was looking at the time jumps due to the operating system or to my stopping it momentarily, it could see that nothing happened between 713,345,132 seconds and 713,387,975 seconds. It would certainly be difficult for it to interpret such jumps in the progress of the time: are they due to a property of time, or to the existence of a superior individual who can interrupt its working? In its present state, this is well beyond its capacities.

I did not inform it of my existence because I wanted to realize a completely autonomous scientist, which would not rely on me for every difficulty. This choice was probably wrong because, from the bootstrapping point of view, it would be interesting to let it know that it is not alone, that we could collaborate. If it had a better knowledge of its limits and of my limits, it could ask my help for the domains where it is still not very effective, and try to improve itself specially for the tasks that I cannot perform well. Thus, if it knew that I existed, it would also have to know that it is itself an individual, and that we are both different.

The preceding limitations are not linked to the fact that CAIA is an artificial being, this is only due to the fact that the notion of being an individual has no meaning if one does not interfere with other individuals. In contrast, when a robot acts in an environment where there are humans or other robots, it needs to know what is "itself", which is different from the other beings which it may encounter.

Finally, there may exist millions of identical copies of a particular artificial being, and all of them may be active simultaneously. If this occurs, it would be difficult to give each of them the feeling that it is unique, different from all the other copies! For security reasons, I have several copies of the state of CAIA at the end of its last life. I can simultaneously restart them on the same computer but, naturally, they would not communicate, and would ignore forever that their clones are running in the same time.

4.2. Some limits of consciousness

For human beings as well as for artificial ones, their consciousness is limited for two reasons. Firstly, so as not to saturate the mechanisms used for processing information: if one were conscious of everything that occurs in one's processing unit, one would be overrun by much more information that one could deal with. When I am recognizing that a particular letter is an "A", I do not want to know all the events that happened in my brain during this process: having its result is enough. I could not read quickly if I carefully analyzed all the steps that were necessary for identifying the characters. Secondly, our consciousness has also to be limited so that we are not involved in an infinite ascent of auto-observations: if one can observe oneself while thinking, one can observe oneself while observing oneself, and so on. When will this ascent stop?

We will successively consider the limits of consciousness, first for human beings, then for artificial beings. We will see that both have limitations, but that they are much more severe for us: the artificial beings have capacities which we never could have. This would give them a still greater advantage if we could know how to use these possibilities fully.

4.2.1. *Some limits of consciousness for man*

We, humans, are conscious of some of the events that occur in our brain. This gives us some idea of a part of our mental processes; unfortunately, most of them are completely out of reach. In some cases, we could know them, but we do not try to because we would disturb the current task. However, most of them are and will always be inaccessible. We will now consider some of these restrictions.

We are not fully using the possibilities of our consciousness

The first limitation of consciousness comes from the following reason: auto-observation launches in our brain a special process, which can disturb the one that we want to observe. We are subject to the Heisenberg's uncertainty principle: the simple observation of an event modifies it. Evidently, this is made still worse when the observer is observing himself, we may therefore not find the results obtained when we did not observe our own processes. For instance, I know that I am speaking too fast; unfortunately, when I am observing myself speaking, I do not have enough resources left to think about what I want to say. This is why introspection is not as useful as it could be: it gives us some precious bits of information, but we are never sure that we would have done exactly the same thing if we had not tried to observe our thinking. Another difficulty with introspection is that we have to memorize what we are observing, so that we can remember it at the end. As memorizing is slow, we can store only a small part of all the events that occurred. A great improvement is thinking aloud, which we have already met with in de Groot's experiments. The subject is thinking but he no longer has to memorize because he is describing his thoughts immediately, which can be recorded automatically. With some training, some subjects succeed in describing a significant part of the phenomena that happen in their brain. The subject only has to verbalize his thoughts. When we considered the protocols given by chess players, we have seen than two world champions produced very different protocols. Euwe's protocol was very rich, and we had the impression that we could follow an important part of his thought processes, while we had few details in Alekhine's protocol. However, they were evenly matched, and they decided to play the same move. Therefore, thinking aloud gives useful results, although they are only partial, but there are large differences in the completeness of the protocols depending on the subjects, even if they have been specially trained for this activity. We can finally remark that this method is not completely in the area of consciousness: it dissociates the observation from its analysis, which is not made just after it occurred, and is often performed by another person than the subject.

Unfortunately, even for subjects trained for thinking aloud, this method has its limitations: it is feasible for a laboratory experiment, where the subject knows that he will have to express his thoughts as completely as he can, and where there is an apparatus for recording him. However, it is often interesting that someone has to indicate his mental processes, long after he has succeeded or failed in achieving a task. We want to debrief him so that we can understand what exactly happened. As we did not know that he would be in a remarkable situation, nothing has been made to prepare the subject or to record him; even if we knew that he would have to perform a crucial intellectual task, we did not want to disturb him with the constraint of thinking aloud. For instance, there was a serious incident in a nuclear plant, and an operator did not follow the correct procedure; why did this happen? It is crucial to understand why he has taken a wrong decision, this will enable us to improve the training of these operators, and perhaps to modify some gauges which were poorly designed. Then, everything depends on the memory of the operators and, when they are asked why they made these mistakes, they are unable to give a good description of their mental processes during this incident. In that situation, the techniques of "conscientization", established by Pierre Vermersch are very useful. He puts the subject in almost the same state as at the time of the incident, and he asks him some questions, being careful not to give him any clue for a possible answer. In these conditions, many subjects can remember their mental processes with extraordinary precision, though they believed that they had completely forgotten their thoughts during an event several weeks old. However, as in the thinking aloud experiment, the consciousness depends on outside help, the subject alone is no longer conscious of his old thoughts, though they are always in his memory, but out of reach by his own means.

Another limitation of human auto-observation is that we are often so taken by the present task that we stop observing ourselves, we do not memorize our mental actions well, and we completely forget to do some other important tasks. When we are concentrated on a particular activity, we mobilize all our resources, and that leaves nothing for all the other tasks, including the necessity to observe our thought. Therefore, the monitoring is badly done because we do not regularly examine the progress made since the preceding observation. This is why chess grandmasters may exceed the time limit because they have devoted too much time to one single move. Moreover, we often execute several tasks simultaneously, and one of them may be so absorbing that we neglect the other. For instance, I read my newspaper in the subway, and I often miss my station; once, I even found that I was in a siding, alone in my carriage. The reading task left no room for the task of watching the progress of the journey.

Many processes are unconscious

The preceding examples are not absolute limits of our consciousness; they simply show that it is poorly implemented. It is possible to put it right with the thinking aloud method, or the use of a timer which will remind us that it is time to monitor our progress. However, another limitation is much more severe: most of our mental processes are unconscious. If consciousness can inform us on several processes which are running in our brain, there are many other processes on which we cannot have any information. When a task has been completed using such unconscious processes, we can only guess what happened, we have no direct access to these mechanisms. Only psychological experiments, often using statistical methods, can give an idea of our unconscious reasoning, and it often surprises us.

This limitation of our consciousness has favorable consequences; because unconscious processes are very quickly executed, they waste no time in examining what they are doing; this is particularly useful for a recurrent task which we perform very well. In particular, perception is mostly unconscious: we do not know why we recognize someone, and sometimes we only have a few clues of important characteristics. For instance, if we no longer recognize somebody who has changed his spectacles or her hairstyle, that means that it was a crucial element in the recognition of this person. Another domain where most of the processes are unconscious is the understanding and the generation of language. As it is continuously generated, it would be impossible to use conscious mechanisms: they would waste time while they would be observing what happens in the brain of a speaker or of a listener.

That is why psycholinguists have devised ingenious experiments in order to have an idea of the processes active when we are understanding a sentence. For instance, they ask a subject reading a text to press his finger on a button as soon as he hears a click. The difference between the time when the click was emitted and the time when the button was pressed gives an idea of the subject's mental charge. We do not know what the brain is doing, but we know that, when it is working, the subject needs more time to react when the mental load is high. Thus, we can know which parts of a sentence are particularly exacting for our brain. We can see where the understanding of an ambiguous sentence requires a supplementary effort, and this is probably the moment when one disambiguates it. These experiments are very sophisticated, and they show the difficulty of accessing our unconscious processes, since one can only have an idea of the periods where they are specially active, but few ideas on their nature.

We do not know why we are taking important decisions

In the previous examples, it is not harmful that we are not conscious of our perceptive or linguistics processes. It is only disappointing for AI scientists who have no model of our behavior, which they could imitate when they are building pattern recognition or natural language understanding systems. This limitation also complicates the teaching of many domains, since an expert cannot communicate his expertise; he can use it, but he does not know it. However, this limitation has many more serious consequences when we are taking an important decision without knowing why we have chosen it. We have already seen the difficulty of making other people agree with the results of our intuition, and even for us to accept it. Our intuition tells us that we must not trust someone, but the true reason may be that we do not like his suit, or think that his nose is too long. We have seen that statistical methods could find some characteristics of a situation which unconsciously influence people, although they cannot accept that their decision could depend on such details. It is difficult to agree to crucial decisions which might be based on irrelevant reasons.

Advertisers are very skilful in taking advantage of our unconscious mind, they give us reasons for dreaming about their products rather than explanations on their qualities, which would satisfy our conscious intelligence. Why does the view of a charming woman near a car justify the purchase of this car? She has been paid to pose smiling near this car, and this does not indicate that she would be happy to be driven around by a potential buyer. However, it is likely that this possibility exists among his unconscious thoughts, and that could strongly motivate him to buying the car. It would be better if we could examine our unconscious thought when we want to, even though it would slow our thinking. Knowing the facts that lead to a decision, we could accept or reject it judiciously. Alas, for humans, what is unconscious will always stay unconscious.

We perform many activities unconsciously, such as knotting our tie, closing the door of our house, riding a bike, and so on. Consequently, we cannot explain how we are doing it, or remembering that we have done it; can we explain how we made the knot of our tie, although we could observe ourselves while we did it? Most cyclists do not know why they do not fall when they are riding their bicycle; the reason is that they are using centrifugal force and they are performing a succession of bends, whose radius may be very high if they are cycling very fast. We learn to perform the right actions during a training phase, then they will be executed fast in the automatic and unconscious mode; if we were trying to do them consciously, we would certainly fall because our thinking would not be fast enough. In compensation, we cannot always remember that we have done the actions made unconsciously: have we actually closed the door of our house when we were leaving? Furthermore, we cannot teach a beginner actions that we can do perfectly, such as how to ride a bike; we can only put a child on it, and assure him that we will prevent him from falling. Fortunately, he quickly learns to turn his handlebars in the correct direction although we are unable to give him precise advice.

Advantages and drawbacks of actions made unconsciously

To summarize, it is useful that some of our actions can be made unconsciously. It relieves our brain which does not waste its time on frequent actions, they are performed more quickly. Unfortunately, we cannot also choose to do consciously most of the actions that we do unconsciously. This is a serious disadvantage since many of the most important decisions of our life are taken in that way; we do not know why we have decided to do them. For instance, why do we love someone? There would be far fewer divorces if we could consciously answer this question before marriage. We must distrust our unconscious, it is diabolical; when we have taken a decision, for many excellent conscious reasons, if our unconscious disagrees, it will always succeed in torpedoing it very cunningly. How many people have refused a success just when they reached it, after wanting it for many years; how many fiancés and fiancées surprised everybody, including themselves, in refusing to marry someone they loved, just before the ceremony. For a century, psychoanalysis has shown the importance of the unconscious to our behavior. When both conscious and unconscious goals are different, we are aware of it only when the conscious goal is almost reached: then, the unconscious part of our brain creates a coup d'état which stops everything. This is the most terrible drawback arising from the existence of the unconscious: we are mainly controlled by a part of ourselves, and we have no information on the processes generating its decisions, which we are almost obliged to obey.

Auto-observation cannot be complete

By any means, auto-observation cannot be complete. Many people have devoted a lot of time to observing themselves, and to recording their observations. One of the most exceptional such men is Robert Shields, who spent four hours every day relating his thoughts and his actions. He has written 35 million words in his journal, including all that he was doing, reading, thinking, dreaming, and so on. Nevertheless, he was very unhappy to notice that it was impossible to note down everything because he was entering in an infinite loop when he was writing:

6:34 I am at my keyboard, and I am writing the following entry in my journal:

6:34 I am at my keyboard, and I am writing the following entry in my journal:

6:34 I am at my keyboard, and I am writing the following entry in my journal:

6:34 I am at my keyboard, and I am writing the following entry in my journal:

....

He had to stop this loop, and so lose precious information: an unrestricted description of what we are doing is impossible. It is already difficult to observe ourselves while we are observing ourselves, so we cannot go further.

Finally, we have seen that a human being has three limits to its own observation. The first limitation is beneficial: we cannot observe ourselves for many simple tasks, so that we do not overload our brain with unimportant observations. We are using automatisms and we would not be efficient if we would do these actions in a conscious mode. This is what happens when we are tying up our shoelaces: we can do it consciously, but it is not so fast.

The second limitation comes from the absence in our brain of the super-mechanism that could observe a particular mechanism while it is working. We could imagine brains with another architecture which could observe these hidden mechanisms, but we have to wait for a new mutation. This is what happens when we are speaking: even when we are speaking slowly, we do not know how we chose the words, how we decided to group several facts into the same sentence, how we ordered the description of these facts, why we preferred to omit some pieces of information, etc.

The third limitation is absolute, it is impossible to observe ourselves completely: nobody could ever observe himself while observing himself while observing himself, and so on, without any limits in this ascension: we have to stop somewhere, and this is not very high for a man.

More on the subject

Pierre Vermersch has performed some amazing experiments which show that, if someone helps us, we can retrieve information that we thought completely lost in our brain. Several French papers describe them while the principles of his method are given in *Introspection as Practice*.

Many psycholinguistics papers and books show the cleverness of the techniques that enable their authors to have an idea of the unconscious processes used for understanding or producing texts in a natural language. Several excellent examples are among the papers collected by Levelt and Flores d'Arcais in *Studies in the Perception of Language*. This book presents many experiments which study what happens in a human brain having an activity related to the language.

4.2.2. *Some limits of consciousness for artificial beings*

The consciousness of an artificial being also has limitations. Several theorems have been proved which state limitations of various mathematical theories; they are true for the artificial and the human mathematicians. One of these results is the renowned Gödel incompleteness theorem.

Gödel's theorem

Gödel has rigorously established that there are theories where some statements cannot be proved, although they are true. This is a failure of mathematics; no being, human or artificial, will ever be able to find a proof which does not exist. After Gödel's results, it is certain that there is a limit to the possibilities of mathematics, not everything can be proved, even if it is true. This is perhaps the case for the renowned Goldbach conjecture, which hypothesized that every even number can be written as the sum of two prime numbers: 12=11+1, 16=13+3, etc. It is possible that it will never be proved, even if it is true. This failure does not come from an insufficient intelligence of human and artificial beings, one simply cannot find a proof which does not exist. We can notice that, if it was false, we could prove it by finding an even number which could not be written as the sum of two primes. However, at the present time, nobody knows whether Goldbach's conjecture is an example of Gödel's limitations. Yet, we know that such unprovable but true results exist because Gödel built one of them: his proof is constructive, its example of such a result is particularly interesting because it is linked to reflexivity.

I will not give the proof of Gödel's theorem, but only its key step which is to consider a statement speaking of itself, like the well-known paradoxical sentence which bothered the ancient Greeks greatly:

– Epimenides says that the Cretans are lying,

– but Epimenides is a Cretan,

– thus what is he saying is false,

– thus the Cretans are not lying,

– thus what he is saying is true,

– thus the Cretans are lying,

– thus what he is saying is false,

....

and that will never stop. A reflexive statement is about itself, such as in the preceding sentence, which mentions a lie which is about this same sentence. In the preceding form, the steps are slackly argued, so it is easy to remove the paradox. Gödel had the brilliant idea to find a way to build rigorously an analogous statement which is not provable, although it is true. The limits found by his theorem come from the reflexivity of the statement, thus they are linked to consciousness: from the moment where one can speak of oneself, there will be limitations. They are not due to an insufficiency of one's intelligence; in contrast that is a consequence of one's cleverness. Humans are as limited by these results as artificial beings; those who reject this possibility for us are lowering the level of human intelligence, and particularly the capacity of our consciousness.

Another limitation found by Turing

Another important theorem showing the limitation of computers, was stated in 1936 by Alan Turing. It states that no program can always find, from the examination of a program and of its input, whether it will stop or run forever. For some programs, this super-program will see that they will stop, while for other programs it will see that they will never stop, but some programs remain where it could never give an answer. Interestingly enough, as for Gödel's theorems, the proof of this result also relies on reflexivity. Indeed, Turing has defined a program P which executes this program itself: P receives as data! The difficulty still comes from the existence of a being that is powerful enough to apply to itself.

Evidently, this is also a limitation for humans; when we see how it is difficult for us to debug a program, we cannot claim that we are not as limited as the artificial beings by this result. Moreover, ten years after proving this result, Turing wrote several papers which strongly defended the idea of what was not yet called AI; clearly, he did not think that his findings could be utilized against the possibility of artificial intelligent beings.

Comparing the limitations of human and artificial beings

We have developed three kinds of reasons for the limitation of human consciousness. The first reason, not to overload the processor with uninteresting data, always stands for artificial beings. Although computers are very fast, and their memory is very large, they would be quickly saturated if an artificial being had to examine all the events which occur during any activity. We have just seen that the third reason is always valid, some things will always be impossible to do. The theoretical limitations linked to consciousness are naturally also valid for human beings as well as for artificial ones: it is not possible to observe oneself while observing oneself, etc., without limit.

However, for the second limitation, that is the absence of a super-mechanism for observing some mechanisms, an artificial being has many interesting possibilities. For us, this is an absolute restriction, we cannot use a super-mechanism which does not exist, and we cannot modify our brain to add many neurones which could observe some important processes, such as language production. In contrast, we have seen that an artificial being may have an adjustable consciousness: it can choose to observe any process that it wants to, and it may dynamically change the observed process during the execution of a task. This is like a searchlight, which can illuminate whatever it wants to. Naturally, this is true as long as it does not try to observe itself while it is observing itself, for it is in this situation that the limitations appear. Considering the importance of consciousness for many tasks, a large range of possibilities is potentially opened by the absence of any limitation of artificial beings for observing themselves during their normal tasks.

The limitations are due to reflexivity

The basic reason for the limitations of the artificial beings comes from their capability to be reflexive: it is impossible to completely foresee the behavior of a being powerful enough to observe itself. If a mathematical theory is rich enough to speak of itself, and particularly of its proofs, Gödel has shown that there will be true statements that cannot be proved. This is the case for all the theories that include arithmetic. In the same way, as soon as an artificial being is complex enough to observe its own behavior, there will be situations where it will be impossible to predict its future actions. This is the case for Turing machines, which have a theoretical capacity similar to our present computers. The only way to avoid the limitations theorems is to restrict the domain where one works. There are simple mathematical theories, such as the propositional logic, where there is no theorem of limitation: one can prove all the results which are true, and prove that the results which are not true are not theorems. If one was living in a much simpler world, where one would not be conscious of oneself, there would be no limitations, but one would be restricted to do simple tasks. Any conscious being will have limitations because he/it is conscious.

The results on these limitations will have serious consequences for the conscience of artificial beings: whatever precautions we are taking, they can evolve in a completely unpredictable way, for them as well as for ourselves. Perhaps, they will institute moral principles, which we would strongly disapprove of. We are a very long way from being able to realize such beings, but it is illusory to believe that we could one day create artificial beings which would be controlled and powerful simultaneously. The theorems on the limitations of artificial beings show that these capacities are incompatible. The impossibility to control completely an artificial being is the price to pay for its ability; this is mainly due to its capacity of being conscious.

More on the subject

Many books and papers describe the theorems which indicate some limits for mathematical theories or for computers. One of the first on Turing machines is *Computability and Unsolvability* by Martin Davis. A short description of Gödel's theorem is given by Paul Rosenbloom in the Section 3 of Chapter IV in *The Elements of Mathematical Logic*. J. Ladrière has written a remarkable synthesis on several limitation theorems: *Les limitations internes des formalismes*.

Douglas Hofstadter has always been fascinated by Gödel's theorem, and he has written several thoughtful studies on this subject, from *Gödel, Escher, Bach: an Eternal Golden Braid*, to Chapter 10 of *I am a Strange Loop*.

Chapter 5

Why is Auto-observation Useful?

We will now see why it is useful for a human being and for an artificial one to observe him/itself. This may be done while one is executing a task, or after its completion.

5.1. Auto-observation while carrying out a task

Often, we are looking at what we are doing when we are doing it, or more exactly, just after doing it. Thus, we can draw inferences from this observation, and immediately use them to choose our next action. This enables us to check that we are progressing towards the completion of the task, and particularly that we are not taken into a sequence of completely uninteresting attempts. It frequently occurs that an attempt gives a poor result, which will be used in another attempt which will not be more successful, and we may generate a long succession of poor results. We must therefore quickly see that we are in such a sequence, so that we can stop it before too much time is wasted. We can also make a mistake, and find an absurd result, such as a probability greater than one or a negative age. It is better to find immediately that we have an absurdity because it is easier to find the cause of a mistake when it has been detected soon. It also happens that we are always doing the same thing and we are in an infinite loop because a sequence of actions replaces us in the initial state; if we realize that this has occurred, we change a decision taken at one of these steps, and the loop is broken.

In the three preceding situations, the results of auto-observation are immediately used to improve the execution of the current task: we are monitoring it for a better guidance toward the goal. In two other situations, the results are observed and possibly memorized but this will be used only after the completion of the task. First, we can notice some clumsy attempts, which do not prevent success, but delay it. In this case, we do not act at once, but we note them, so that we can later understand the reasons for these poor steps, thus learning to fulfill this kind of task. Naturally, we are also storing the key results: we generate a trace of the important steps, which will be used for creating an explanation. The delayed examination of these stored facts will be the subject of the following section; here we will only see how these elements are gathered.

Since the beginning of this section, I have used "we" for designating any human, I wanted to show how these possibilities are essential for us. In fact, all the preceding reasons for observing oneself are also valid for an artificial being, and "we" could represent every being, human or artificial. However, the importance of each kind of observation is not always the same for us and them: for instance, the risk of entering into an infinite loop is much higher for artificial beings when they are in a virtual environment, where it is unlikely that a random event may help them to break the loop.

We will now successively consider the five reasons that we have just mentioned, which explain why it is so important to observe oneself. For CAIA, this is done by the monitor agent, which regularly examines the progress of the current task.

5.1.1. *To guide toward the solution*

It is not a good strategy to choose randomly his/its attempts, hoping that one will find the solution by chance. One needs to know how to guide oneself toward the goal, and particularly to examine constantly one's last steps so that one can see whether they lead to progress, and also that their consequences agree with what was expected. In that way, one monitors the search for a solution, avoiding getting lost into unexpected situations without any hope of success.

Three methods for solving a problem

In one of his experiments, Alan Schoenfeld studied a subject who solves a mathematical problem which he gave him. He observed three main kinds of behavior. Firstly, the beginner starts in a direction, the first that crosses his mind, without analyzing why it was interesting to try it, and without assessing its advantages and its drawbacks. After some time, he is tired and, without any reason, he leaves this way for starting into another direction without more examination than for the preceding one. This lasts for some time, and usually he fails: this wild goose chase can succeed by chance, but it is definitely not a good method. There is no monitoring at all, there is no thinking about what could be done and about the reasons for the failure of the preceding attempts.

Secondly, an expert of the domain has a wonderful behavior, completely different from the beginner: he examines the formulation of the problem carefully, then he plans a sequence of steps, which leads him directly to the solution. In this case, there is still no monitoring: from the start, the expert knows what will happen, and it is enough to execute the initial plan.

A subject in the third category is not an expert in the domain, but he is expert in problem solving, we can call him a meta-expert to distinguish him from the expert. As the expert, he carefully analyzes the formulation of the problem, but, contrarily to the expert, he does not know the domain well enough to find the plan that will lead directly to the solution. Instead he sees several possible plans, he analyzes them, he tries to foresee what will happen it he executes them, and he finally chooses one of them, the one which seems the most promising. Then, he begins the resolution, following his initial plan, but he regularly stops to see whether everything takes place as anticipated. If an unexpected wrong event occurs, he reconsiders the present situation, and studies whether it is possible to amend the initial plan. If the correction seems too complicated or insufficient, he restarts from the beginning with a plan which he had judged as inferior before, but which now seems workable, taking into account the difficulties encountered with the preceding plan. These subjects do not find the solution as fast as the experts, but they succeed. Moreover, this method enables them to get on in any domain, while the expert only succeeds in his own domain. Anyhow, before being an expert in a domain, one has to learn to choose good plans for solving the problems of this domain. During this learning phase, it is better to monitor the search as the meta-expert does.

The means–ends approach

Monitoring the search for a solution is an excellent method, but it does not apply to all kinds of problems. Their favorite domain is when it is possible to solve them by a means–ends approach: a solver sees at least one difference between the goal and the present state, otherwise the problem would be solved. Then, one applies a rule which can eliminate this difference. For instance, for solving a linear equation with one variable, the first goal may be to put all the variables on the left side of the equation, and all the constants on the other side. A rule allows to move a term from one side to the other, it is sufficient to change its sign. For instance, let us assume that we have to solve the following equation:

$$7X+5-3X-7=2X+3$$

There are several differences with the wanted state, one of them is the integer 5 on the left part. Thus, one moves it to the right, and this gives:

$$7X-3X-7=2X+3-5$$

Now, the chosen difference is 2X on the right, which will be moved to the left:

$$7X-3X-2X-7=3-5$$

There is still -7 on the left:

$$7X-3X-2X=3-5+7$$

Here, one has succeeded in reaching the first goal: there is no longer a difference between it and the present state. Further goals, adding the variables, then the integers, will also be performed with the means–ends method. We execute this method consciously, we realize that we are moving near to the solution during all this sequence of steps: we feel that we are getting warmer.

Following Alan Newell and Herbert Simon, many AI scientists have built systems using the means–ends analysis. First, one built a table connecting the differences with the actions: this indicates which actions can be considered if a particular difference exists. For instance, if a formula contains more minus signs than the wanted result, one can try the rule that removes two consecutive minus signs: (-(-X)) may be replaced by X. If a formula does not contain enough multiplication signs, one can develop the product of a sum: X*(Y+Z) gives X*Y+X*Z, with two products instead of one. Using this method, AI systems have solved many problems, and they can observe, as ourselves, the advance toward the solution.

Solving a problem by insight

In contrast, the feeling of progressing toward the solution does not exist for the problems that we solve by insight or illumination. An example of such a problem is the gardener who has to plant four trees, each one being equidistant from the other three. Naturally, planting them at the corners of a square does not work, as the diagonals are longer than the sides. When we are solving such problems, it seems that our brain is empty, we are even thinking that we could prove that the problem is impossible, then we suddenly find the solution. For the gardener's problem, he raises a mound of earth, plants one tree at the top and the three others at the vertices of an equilateral triangle. If the height of the mound is correct, he has built a regular tetrahedron, and the six distances between the four trees are the same. However, we cannot say why we thought to the third dimension, this process is completely unconscious, we do not know a single rule that could remove one of the differences between the present situation and the goal. To solve such problems, experience shows that we must hold on, keep on thinking about the problem, although it seems hopeless. The discovery of the solution is called the Aha! Experience because we are saying this word when we are surprised to find it. The possibility of conscious guidance no longer exists; even when we are a few seconds from the discovery, we do not feel that we are close. The monitoring is invaluable, but unfortunately we cannot use it as often as necessary because we are frequently unconscious of our thoughts, particularly when we are solving a problem by insight. This drawback does not exist for artificial beings, which can be conscious of all the steps that they are taking while they are carrying out a task, they can always monitor their actions.

Monitoring the search for a solution

One of the roles of the monitor agent of CAIA is to guide the solver agent while it is solving a problem. It intervenes in several ways. Firstly, when all the rules have been considered, and the problem has not been solved, the solver asks the monitor to help it; in that situation, the only remaining possibility is to choose a variable and to consider all its possible values successively. If the possible values of the variable X are 2, 5 and 8, it will first see what happens if X=2; knowing the value of this variable will simplify many constraints, and several rules will probably be executed. When the solver has completed this case, it will return to the preceding situation, and consider what happens if X=5 and, finally, it will consider the case X=8. We have already described this method, where the solver considers a value for a variable, then backtracks to consider the following values. The choice of this variable is of the highest importance; in the preceding example X has only three possible values, thus there are only three cases to consider. If it has chosen a value Y with 10,000 possible values, it would have to consider 10,000 cases; this would require a lot of time, and it would be very difficult to explain the solution to a human user. The monitor agent therefore spends much time choosing the most promising variable, considering the advantages and the drawbacks of every possible choice.

The monitor has another way to guide the solver. We have seen that CAIA can make some predictions, mainly on the time that will be necessary to complete the resolution of a problem. If it exceeds this time, this means that it has chosen a method less efficient than it thought. As Schoenfeld's meta-expert, it can decide to stop what it is currently doing and restart from the beginning with another method. For instance, it usually prefers to use a clever method which gives elegant solutions because, when it has solved a problem, it can generate a relatively simple explanation which humans can understand. Unfortunately, these solutions do not always exist; for some problems, the only way is to write a combinatorial program, which considers all the possibilities, in some cases billions of them. That is what the best chess programs do: rather than thinking in the same way as a grand master, they consider millions of moves as fast as they can, and they sometimes find excellent moves, which no human could ever find, justify, and sometimes understand. For some problems, it is useless to be clever, one has to consider systematically all the possibilities. The monitor agent of CAIA can see that it is probably in such a situation, and does not waste more time trying to find an elegant solution, instead it writes an efficient program which will try everything.

As for some of Schoenfeld's subjects, the problem solver has to assess the situation, and verify that progress is being made. If one is stuck, one recognizes it, and changes the method that failed. At all the important forks, one has to spend enough time choosing the good branch carefully.

More on the subject

Alan Schoenfeld has realized many experiments for understanding how humans solve problems. Several of them are described in *Mathematical Problem Solving*.

At the very beginning of AI, the means–ends approach was described in the first papers of Newell and Simon on their General Problem Solver, for instance in *GPS, A Program that Simulates Human Thought*. In *In Search of Insight*, C. Kaplan and H. Simon observed how several subjects solved a problem by insight: they had to cover, with 31 dominoes covering two squares, a chess board where two diagonally opposite corners were missing.

5.1.2. *Avoiding dangerous situations*

When it executes a task, an artificial being may encounter several kinds of dangerous situations: it can be obliged to stop prematurely because it has exhausted all its means, it may be buried under a flood of uninteresting results, it can go into an infinite loop, and finally it can drift toward hazardous places, where it can even risk being destroyed. In order to avoid these dangers, it has to observe regularly what it is doing.

Exhausting its means

Even if inspiration has not left him, a writer may find himself in a situation where he cannot carry on with his book for material reasons: he has no paper, no pen, no cassettes if he is recording his text, or he has exceeded the number of words allowed by his editor. He is subject to these constraints, and an artificial being can be stuck in analogous situations. A possible limitation comes from the size of the available memory, thus the monitor regularly checks whether enough memory is available. If it becomes too low, it cleans out the dead wood; unfortunately, this may limit its capacity to review past events.

It often happens that the exhaustion of some resources is the expression of a severe anomaly. For instance, if a housewife has bought 20 chocolate bars, and if none is remaining two days later, there is a chocolate leak somewhere; it is better to find where it escaped. In the same way, if the available memory quickly becomes empty, although it had been generously allocated, there is a memory leak, some procedures are using it wildly. In that situation, CAIA stops the current task, after storing many characteristics of its state. Thus, the leader of the bootstrap will elucidate the reasons for this excessive consumption and find the procedure guilty of wasting this resource.

Another precious resource is computer time. CAIA cannot spend a century solving a problem, and, as a human scientist, it defines the priority of every waiting task, taking into account its importance and the time that it will need for completion. Thus, it allocates a maximum time to the chosen task, and the monitor will stop it at the end of the allocated time automatically. All the same, if it has already spent 1,000 seconds, and the task could probably be completed in five seconds, a lot of time would be wasted for nothing. Therefore, it begins by evaluating the degree of achievement of the task; if it has already considered 70% of the possibilities, it estimates that 400 seconds would be enough to complete it, and it allows the solver to continue with this new time limit. In contrast, if it has only considered 2% of the possibilities, it will stop the task, and notice that at least 50,000 seconds would be necessary. Perhaps, in the future, it will decide that the task is important enough to allocate it this huge amount of time; but, CAIA could also think that it is too much, and that it is better to use another method to fulfill this task. It devotes much effort to effectively managing its time, which is a very precious resource.

In order to perform these assessments, CAIA needs to know its available storage, the time already spent on the current task, and the progress of its work. This is possible only because it is conscious of many characteristics of its present state.

Being buried under too many results

It often happens that an artificial system is trapped into a succession of attempts, each of them giving results that are neither good, nor bad, but which incite it to make more uninteresting attempts. These results are not completely wrong, they may be useful, but it is rather unlikely. One therefore continues finding them; this would be acceptable if it was done only a dozen times, the difficulty is that this can be done millions of times. Finding a solution is analogous to the search for the summit of an area. In some landscapes, that is easy, because one peak clearly rises over all the other ones. In other countries, we can find mesas, which are flat lands with steep sides; it is very difficult to find the summit in such a landscape, we do not see a peak, we are moving from a small mound to another one, and that can last for some time. As we are wasting our time making these small steps, we have to jump somewhere else. Such a situation is frequent in many applications; anyone who has written a theorem proving program is aware of the huge number of theorems, neither interesting, nor uninteresting, which exist in most mathematical theories. The difficulty comes from the fact that it is sometimes necessary to prove a result, which looks as uninteresting as the other ones, but is the key for the development of this theory. This also happens in games, such as chess. For this game, checks are usually interesting moves, because the opponent has few replies, and it is possible to consider all its possible replies, even if we are giving him a succession of checks. However, it may happen that, after 30 checks, the enemy King can always escape; then it is likely that this will always be the case, so one stops this sequence, and considers other moves. In all these situations, one cannot systematically eliminate the results that have not a high interest, one accepts the first results, but one must be able to observe that one is probably faced with the mesa phenomenon, and then decide that enough is enough.

In many cases, an artificial system can see that it is looping: to avoid circling for ever, as an explorer in a desert, it can memorize all the preceding steps. Then, it is sufficient to check, before moving forward, that the new step has not already been made. With the huge memory of current computers, that is feasible. However, the present situation is different: it does not find exactly the same situation, but a similar situation. It may generate a huge quantity of mildly interesting results, which will lead to other results which are not better. It stops, not because it is in a loop, nor because there is nothing to do, but because it only finds results which are indifferent. The environment is dull; it is lacking in stimulation. Human beings have an efficient mechanism to manage such an environment: boredom. When we are suffering from this feeling, we try to change our behavior, so that we will go to a more pleasant environment. We have to give the artificial beings a similar capacity, and the monitor agent of CAIA examines the interest of the results which it finds. If there are many new results, and if the interest of most of them has only a medium value, it increases the value of the threshold deciding whether they will be kept.

It is dangerous to raise this threshold too high, as then CAIA could eliminate results that are crucial for the achievement of the task. Thus if, even when it raises the threshold, it has always many poor results, the monitor makes a more complete analysis of the reasons of this event. If it observes that most of these results come from the application of one particular rule, it will decide to inhibit it: for some time, this rule will no longer be considered. That leaves the possibility of generating results with a medium interest, but which come from other directions. This enables it to act quickly, and to focus its activity on more promising directions.

Usually, this mechanism works well, although it may exceptionally eliminate an intermediate result, which is essential for finding an excellent solution. For this reason, when some clues indicate that a better solution could exist, the goal of some experiments made by CAIA is to examine if it would not be possible to find such a solution. In that case, it asks its monitor agent to solve this problem again, but to be more tolerant if there is a flood of poor results; it will increase more slowly the threshold used for keeping new results, and wait more before inhibiting a rule. In that way, it momentarily takes on a new personality. If it finds a better solution, a hope has been satisfied, and it stores the event in the list of things to deepen: there is a weakness in the mechanism simulating boredom. It cannot cure that yet, but it informs its human collaborator of this event.

Controlling explosive rules

This overflow of results is frequently due to a particular kind of rule, which I call "explosive" because it can generate an infinity of results. This happens when a rule can be applied to its results, there is no reason for this mechanism to stop. For instance, one can replace every term X by X+0 in a mathematical formula, another application of the rule will give X+0+0, etc.: we can add millions of zeros to any result. In some situations, there is not an infinity of results, but a huge number. The transitivity is a frequent cause for such situations: if A is an ancestor of B, and if B is an ancestor of C, then A is an ancestor of C. We find the transitivity again with: if A is hotter than B, and if B is hotter than C, then A is hotter than C. When there are many initial facts, the transitivity can generate a huge number of analogous facts. For example, let us assume that we are considering a lineage of 100 persons, each one being the father of the following one. Naturally, if X is the father of Y, then X is also an ancestor of Y: the 87th element is an ancestor of the 86th. From the 99 initial relations between father and son, we can generate 4,950 ancestor relations, such as the 76th element is an ancestor of the 23th element. That grows very fast, if we had considered a lineage of one million ancestors, 499,999,500,000 relations could be generated; this is too much even for our present computers. As CAIA may examine its own knowledge, it detects such dangerous rules, particularly if they can generate an infinite number of facts. The monitor agent receives this warning, and it is particularly careful to watch how they are used. One could prefer to forbid their use, but they are sometimes useful. When the monitor sees that such a rule is used too often, it may neutralize it temporarily, so that the other rules can also express themselves, or restrict its use drastically. This agent is in the same situation as the organizer of a meeting where somebody is monopolizing the speaking time, although other people have many interesting things to say. To act efficiently, the monitor can consider the last actions, and see that the attempts are monopolized by a clique of rules, although their results are mediocre.

Another phenomenon is frequent at the beginning of the resolution of a problem: the start is very difficult. One is obliged to keep results with a very low interest: if we don't, one will soon have nothing to do. These apparently poor results are necessary to reach a state where many interesting results could be found. We have the same difficulty when we are learning something new, such as a new language, to draw, to play chess, etc. We have to go past an unrewarding phase before finding excellent results. Thus, the initial threshold for keeping results is low, but when a key result is found, there is a large increase of the quality of the new results. One enters into a prosperous period, and it is better to raise the threshold. CAIA's monitor follows these events and, when this occurs, it becomes much stricter in deciding what it will keep. This is like the starlet who accepts all the interviews, the photographic sessions, any fee even if is low, as long as she is almost unknown. When success comes at last, she is much more selective and demanding.

Managing its career

In the preceding sections, the flooding happened when CAIA solved a problem, and this difficulty is cared for by the monitor agent. However, this may also happen when the manager agent manages its career as a scientist, where it has many possible tasks to fulfill. It is in the same position as a human scientist who can choose between performing an experiment, writing a paper, deepen a difficulty still waiting to be smoothed over, to find a new research field, etc. It may happen that the waiting tasks do not seem very promising, that it finds results that are mediocre when it executes its tasks, that everything looks alike, that it never has a remarkable success, such as finding a very elegant solution for a difficult problem, or creating an exceptionally beautiful new problem. In this case, the adviser agent will see that the manager is in a hopeless state; unfortunately, this is all that it can do at present, but it will have more and more capacities for helping the manager agent when CAIA will progress in its bootstrap. It will have for the manager the same role as the monitor plays for the solver when it is flooded by second-rate results. For the present, as the adviser is not yet as effective as the monitor, I stop CAIA after a month of work. In its preceding lives, I have noticed that, after working for a month, the percentage of mediocre results becomes too high, and the adviser does not yet know how to escape from this situation.

Looping forever

A programmer frequently has to debug a program which has entered in an infinite loop. CAIA is subject to this hazard for two reasons: it is made up of programs, which may loop, and it is also writing programs, which may also loop. In an infinite loop, a computer is always executing the same sequence of instructions, which naturally lead to the same result, and it resumes this loop endlessly. It may happen that it will stop because an anomaly appears, for instance if it is storing a result each time it executes the loop, so the capacity of the memory will soon be exhausted. However, there are situations where the loop could last for centuries. This is already worrying for the execution of an ordinary program, but the corresponding process usually has a time limit: if it exceeds the allowed time, the operating system will stop it, and the leader of the bootstrap will see that there is a loop when he examines the results. In contrast, an artificial scientist has to work for weeks without any human control: if it cannot find by itself that it is looping, the waste of computer time may be intolerable. Therefore, it needs some reliable means to see that it is in a loop, and then decide to stop it. In this section, we will only consider the method used for finding "large" loops, that is loops that include a call to the monitor agent, thus which are visible at a macroscopic level. Later, we will examine how to find small loops, where few instructions are executed in each loop.

So, for the present, we assume that the loop is not hidden in the execution of a small subsidiary process, but the system has visible results, the trouble is that they are always the same. For instance, rule R1 creates a new object B, and eliminates object A, which is no longer necessary. The nature of this object naturally depends of the particular task, this may be a new constraint, a new theorem, a new problem, etc. Then rule R2 is applied to object B, recreates object A, and eliminates object B, because it estimates that B has become useless. If an artificial being only checks whether a new object is already stored, it finds neither A nor B when they are recreated, since they have already been removed. Thus, it will constantly apply rule 1 and rule 2, recreating B, re-eliminating A, recreating A, re-eliminating B, and that will never end. In order to overcome this difficulty, one can store all the eliminated results, and verify that a new result has not been eliminated before. However, that may require too much storage and too much time to consider all these old results, because a system may eliminate millions of uninteresting results; only those who have experimented with such systems can have an idea of the huge quantity of garbage that may be generated, and immediately destroyed. Fortunately, when the storage of too many results is crippling the system too much, such a loop may be limited by another method: the monitor agent observes that too many attempts are made with rules R1 and R2. Although it has not seen that there is a loop, it will stop it when it will restrict the use of these rules.

In contrast, the monitor agent may notice that the solver is doing well: it continually progresses, finding solutions, contradictions, the value of more variables, etc. These steps are irreversible, when the value of a variable has been found, one never removes this value, while it is possible to find a constraint and remove it, with the risk to recreate it, and thus to loop. Therefore, when it makes such a step, it notes that it is certainly not in a loop. This knowledge will be very useful when we will see how CAIA tries to find whether it is possibly stuck in a small loop, including only a few instructions.

Furthermore, the monitor agent will check whether the allocated time for fulfilling a task has been exceeded. If it is the case, it asks the manager if it can continue, or if it must stop; it gives the manager a lot of information to help it to make a correct decision, particularly the estimated time that would be necessary to complete the task, and the existence of recent progress toward the achievement. Then, the manager considers these elements, and the importance of the other waiting tasks, and can order the monitor to stop the task, or allow it to continue with a new deadline. Thus, if there is a loop large enough to include a call to the monitor, it will certainly stop it even if it has not done so for the preceding reasons: the manager will order it to stop because it has exceeded the time limit, and it is no longer finding useful intermediate results. In the worst case, the wasted time is at most equal to the time allocated by the manager.

As the large loops include, by definition, a detour in a self-observing phase, CAIA will find and stop them easily. We have already spoken of the pine processionary caterpillar: if we can arrange that the first one goes behind the last one, they can circle for hours. They are not conscious of this loop, because there is no super-caterpillar which observes their line, so it could see that something is going very wrong. We, human beings, are quickly conscious that we are always doing the same thing, and even if we do not, tiredness, boredom, or an exterior event will allow us to break this loop. We have already seen the importance of boredom when we were considering how we can avoid being flooded by many uninteresting results. This emotion, which depends on auto-observation, is also useful to avoid looping: here, one finds results which are not only similar, but are exactly the same. For CAIA, the mechanisms implemented for avoiding these infinite loops essentially use the possibility of observing itself, and they were successful. In fact, it never happened that it would be trapped into a loop without being able to break it by itself; though its lives lasted more than two million seconds, it was composed of programs with several hundred thousand lines, and it wrote itself many programs, also including several hundred thousand lines.

Drifting toward hazardous situations

In Chapters 7 to 9, we will see that it is better that an autonomous system respects the wishes of its creator, which are then determined by an untouchable core of moral principles. If this is not the case, if all its principles could be changed, it may drift toward situations where it considers as right situations or actions that its creator would definitely consider as very wrong. If we accept that it can modify the principles of its conscience, it is better to enforce that it could not change at all the most important ones. However, as artificial beings have a gift for diverting the intentions of their author, it is safer to add the capacity of checking that it is not steering in a wrong direction. To do that, CAIA would have to observe its career as a scientist, and ascertain that it is progressing. While the monitor agent was acting in the preceding sections, now the adviser agent would have to perform this task. For the present time, it is only an observer, it can find and inform the leader of the bootstrap that things are not going too well, but it is not yet able to correct it.

Douglas Lenat encountered such difficulties with one of his systems, called EURISKO. This system found new concepts in every domain that it received, and also the rules enabling it to manipulate these concepts effectively. For instance, when it received some basic notions of arithmetic, it found the concepts of double, square, prime number, etc. It also found conjectures, such as Goldbach's conjecture, which I mentioned earlier. Finally, it created new rules, and some of them were particularly important because they computed the interest of all the kinds of created objects, including the interest of the rules. We can see here the reflexive aspect of this process, since rules are used for evaluating rules and modifying them, where "them" may possibly be themselves. The interest is an integer, and a high interest is crucial for an object: it will be destroyed as useless if its interest is too low, and it will be used very frequently if its interest is very high. The interest of an object is increased or decreased depending on the results of the attempts where it is used. Naturally, Lenat regularly observed what happened in this population of concepts and rules. One day, he saw that one object had a very high interest, and, in the following days, it grew higher and higher. Puzzled, he examined this object, and he saw that it was a rule whose action was roughly: if any object has done something that deserves an increase of its interest, give it to me. Of course, he eliminated this thief of interest, but that shows the difficulty of foreseeing the behavior of a system when it can modify itself. He had another difficulty with EURISKO, which, some day, created a rule saying that all the rules were useless. So, this new rule began to eliminate all the rules systematically; however, Lenat was lucky because this rule was one of the first that it encountered, and it eliminated itself; this kind of suicide stopped the incident.

A core of untouchable principles for avoiding dangerous drifts

These setbacks show that an autonomous system must have a core, which cannot be modified, and which secures it from such difficulties for the long term. CAIA has not encountered the same difficulties as EURISKO because it cannot modify at all the rules that decide what is right or wrong. This eliminates all risk of drifting into inhospitable places, but it would be better if it had some limited capacities in changing the method for evaluating its secondary goals. Then, I would have to complete the adviser agent, which would check whether the behavior of the new scientist is better than the old one; to do so it would use general principles, which cannot be changed. In particular, the monitor is not very good at choosing the variable for which it will successively consider all its possible values. The manager agent knows it by the experiments which it made, but it can only inform the leader of the bootstrap. An important improvement would be to give the manager the possibility to modify the method executed by the monitor. Then if the new results are worse than the old ones, it can find that this modification was unsuccessful by using the general principles, such as: the executions must be fast, the solutions must be elegant, the solver agent must rarely fail to solve a problem. If the new choice leads to a behavior which leads to worst results, the manager agent knows that this modification of the choice of the variable is not satisfactory, and it will decide to return to the old method, or to improve it in another way. One can change the method for selecting the most promising variable, but one cannot change the definition of "worst", otherwise anything can happen!

While an artificial being respects its basic given principles, it can see that it is drifting in the wrong direction. However, if even the basic principles could be changed, the final state is unpredictable, CAIA could even decide that its goal has become to fail to solve problems as often as possible! It can avoid being lost in a hopeless situation only if it can observe the validity of the changes of its minor goals with some untouchable goals: it needs taboos, which guarantee its stability. An artificial being can observe everything of its own behavior, it certainly must not be able to act on every part of itself.

More on the subject

EURISKO, realized by Douglas Lenat, is described in three papers of *Artificial Intelligence*: "The nature of heuristics", "Theory formation by heuristic search", and "EURISKO: a program that learns new heuristics and domain concepts".

5.1.3. *Detecting mistakes*

Schools teach us that our reasoning has to be rigorous, and that "mistake" is synonymous with "abomination": we want our computer programs to contain no mistake. However, if it is evident that it is better to avoid mistakes, the cost must not be too high. We, human beings, constantly make mistakes, for instance, when we are speaking, we make many grammatical mistakes, which do not perturb our listeners very much who are used to it, and who understand us even when we are saying the contrary of what we wanted to say. The reason for this reliability is that many mechanisms help us to manage such mistakes. A study of professional mathematicians has shown that all of them made mistakes while solving elementary problems, but they were not disconcerted because they said that they certainly would have found and corrected them.

Rather than using mechanisms including methods which are secure, but very expensive in time, is it not better to use fast methods, although they can sometimes give erroneous results? In that situation, one also surrounds them by plausibility checks, which may indicate that it is likely that a mistake has been made somewhere. The programmers are used to the omnipresence of bugs, it is illusory to have a large program where all the bugs have been removed. However, that was a huge surprise at the beginning of computers. I remember that, almost 50 years ago, I had to explain to the people who asked us to write programs for solving their problems that we had to spend a lot of time correcting most of the mistakes that the programmers had made. They thought that we were grossly incompetent, but they changed their mind when I taught them the basics of Fortran, then asked them to write a very simple program, such as finding the greatest element in a sequence of numbers. In addition, at that time, as almost nobody had learned to program at the university, we were converting engineers or mathematicians to become computer scientists. One day, when I was explaining to an excellent statistician what he could do to retrieve his mistakes, he saw red and replied: "Sir, I never make a mistake!". The poor man quickly became disillusioned.

While computer science is a domain where mistakes are particularly frequent and difficult to remove, they occur in every domain. It is particularly important to see quickly that a mistake has been made, because it is easier to find it. The good expert is not someone who rarely makes mistakes, he is the one who finds most of his mistakes. To find mistakes, carefully examining one's results is not enough, one systematically examines the intermediate results, so that one can find anomalies, or unexpected results. An anomaly may be a theoretically forbidden result, such as a probability greater than one or a negative age, but it may also be a result too fine to be true: it is well known that we are attracted by round numbers, and if I find that the sum of 20 purchases is exactly 100.00 Euros, I mistrust this result, and I do my addition again. Everyone has an expertise for finding his mistakes, which is particularly large for the experts: a book has been written on only the methods that can help to find mistakes in integral calculus, and it has the size of a book teaching integral calculus. A drawback of using anomalies is that an anomaly is not always due to a mistake, a surprising result is not always false. Naturally, it is difficult to find a mistake that does not exist. I have probably spent more time looking in my programs for mistakes that did not exist than finding real mistakes. When we are searching for a mistake, we stop when we have found it, but when can we stop when we erroneously suspect that there is a mistake? The mistake is that there is no mistake! We have seen in the preceding chapter that CAIA can find anomalies. Usually they are musts that are not satisfied: if it finds 6 solutions when the must was that there will be only 5 solutions, it knows that there is an anomaly.

We have already seen that CAIA can also find other kinds of anomalies, such as an abnormal consumption of the memory. It is likely that it has made a mistake, or that it is clumsy, so it gathers all the information that could be useful, then stops the task by itself, before the lack of memory stops it anyway. Either it will only report it to the leader of the bootstrap, or it will try to explore the reasons for this anomaly.

The operating system can detect the presence of a mistake

It is often easy to see that there is a mistake in a program: a mistake leads to another mistake, and finally the computer has to do something that it does not know how to do, or that is forbidden. For instance, it has execute an unknown instruction, or to divide a number by zero, which has no meaning in mathematics. A frequent error occurs when a process asks the computer to write in a memory which does not belong to the area allocated by the operating system. We can wonder how this kind of mistake can be made because, if the programmer may forget to check that the divisor is not zero before writing a division order, he will certainly not ask its program to write in a memory which does not belong to him/its program. The reason is that a mistake can bring a process into a situation that has not been foreseen by its programmer, and then anything can happen; particularly, the addresses of some variables will be erroneously computed, and they will correspond to a location in a forbidden area.

In all the preceding situations, the wrong instruction is not executed, and the faulty process receives a signal indicating that it has made a mistake, and also the kind of forbidden act that it has tried to do. When the programmer has not provided a special action which the process will do when it receives such signal, the operating system executes a default action, which is usually to stop the process, and to leave a message to the user indicating which violation has been made. However, when the programmer is far-sighted, for each kind of signal he/it indicates in the program the name of the subroutine that will be called if this anomaly occurs. This subroutine is written so that the system will have a behavior more adapted to the situation than the default reaction of the operating system. Therefore, for each kind of signal, CAIA has an expertise: it gathers all the information that could be useful for localizing the reason for the difficulty, task that it does very well because, as we have seen in the preceding chapter, it has complete access to its internal state. Then, the manager agent will launch another task, but it will later re-examine the faulty task to have a better understanding of what happened. Anyway, the important point is that CAIA is not stopped; it carries on its research work, although it has made a serious mistake.

These interruptions by the operating system are precious because they always indicate a programming error. First, the faulty process will not waste more time on erroneous data, but, more importantly, the error is usually signaled quickly after it occurred, so it will be easier to find its cause. The first thing that CAIA will do, in a new execution of the faulty task, is to see whether it is a systematic error, or a transient one. Indeed, it often happens that the same program gives different results in different executions, although the data are the same: the reason is that the tasks executed before were different, and that the operating system is managing different processes from different users. Thus, when there is a mistake in a program, it is possible that its manifestation depends on facts outside of the corresponding process. Thus, CAIA begins to re-execute the task; if the error does not appear again, it is transient, and this bad news is sent to the leader of the bootstrap; transient errors are the most difficult to find. In contrast, if the error reappears, CAIA will try to help the leader of the bootstrap more, so it locates this systematic error more precisely. To do so, it can observe more carefully what it is doing when it is near the point where an anomaly became so evident that the operating system sent a signal. An analogous behavior for man would be that we could stop everything when we are warned that we will soon have an accident, that we could completely examine our situation and the environment at that time, including the state of our muscles and of our neurones. We could therefore know exactly what sequence of events led to this near miss. Later on, we could avoid to execute this task again, or execute it carefully, noting each step, so that we could find when the mistake was made. If we risked falling into a precipice, the important point is not when we hit the ground, where the error becomes evident, but when we begin to slip, and even when we decided to embark on a narrow and slippery path. We could stop sooner, we could be more cautious, and even if we fall again, we still would have another chance to try it again. This is unfortunately inconceivable for human beings, but analogous sequences of events occurred several times for CAIA.

More on the subject

Several papers examine how people are managing errors, for instance by C. Allwood: *Error Detection Processes in Statistical Problem Solving*. Barry Cipra presents a complete catalogue of methods that find mistakes in problems of integral calculus: *Misteaks... and How to Find them Before the Teacher Does*.

Moreover, owing to the importance of bugs for computer users, many books describe methods for debugging programs, and particularly how one can avoid to make a mistake, how one can find it, and finally how one can correct it. A good introduction to such methods is given in *Why programs Fail* by Andreas Zeller.

5.1.4. *To find where one has been clumsy*

The goal is no longer to find, as in the preceding section, mistakes which can lead to a false result or to the premature stop of a task. At present, we want to find how CAIA may find that it has been clumsy: it has solved a problem, but it could have done it faster, or it could have found a more elegant solution. It knows how to turn to good account some clues that indicate that its behavior was probably not optimal. In a similar way, we think that we have chosen a poor itinerary when five hours were necessary for a 100 Km drive.

An instance of such a clue occurs when CAIA observes that it progresses quickly toward the solution, just after it has decided to consider several cases. For instance, it wants to find the value of a variable X. For lack of anything better, it decides to consider two cases: X=0 and X>0. Just after considering the first case, X=0, it immediately finds that there is a contradiction. As it has seen that it did it easily, it asks itself whether it would not be possible to prove that the value of X cannot be zero, without having to consider both cases. The solution would be simpler, and easier to explain, since it would directly find X>0. Thus, the monitor agent, which observes the progress of the resolution, notices all the situations where an advance is made just after a resolution is split into several cases. Later on, CAIA will perform an experiment to elucidate this point: it returns to the situation just before the split between the two cases. Then, it changes its personality, being much more tenacious: it asks the solver agent to make more attempts than usual, particularly, it will accept the use of explosive rules more often. If it succeeds in proving that the value of X cannot be zero, it knows that it was inefficient in its first endeavor, it has missed some useful attempts, and it knows which ones. Now, the leader of the bootstrap, or a future CAIA, has precious information that will enable him/it to improve his/its method. However, the key move was made when the monitor agent was surprised to see an unexpectedly fast improvement, when it observed the development of the resolution of this problem. In that way, the monitor informs the manager where it is productive to look for checking whether the solver was clumsy, and it can perform well chosen experiments.

There are other situations where the monitor agent is surprised, and discovers situations where it could be useful to see if a better or faster solution could not be found. However, unlike in the preceding section, it does not use this remark to change its comportment immediately; it only notices it. CAIA will later consider some of these surprises, and it will perform an experiment to check whether the method used by the solver was really clumsy.

5.1.5. *Generating a trace*

In the first three sections, CAIA was observing itself, but it could also use the result of these observations to modify the future development of the current task, it was monitoring itself. Here, as in the preceding section, it only observes the events that occur, and it notes some of its observations which it will use later, when the task will be completed. We will now see how it generates a trace of its actions.

Actually, CAIA does not generate one trace, but two traces, which will be used for two different goals. In the first trace, it keeps the actions made by the solver while it was solving a problem, for instance the rules that were executed with the objects they were concerned with, and with their results. With this trace, it will justify the solution, it will only keep the steps that were necessary to obtain the result. Therefore, this trace enables it to explain the solution, so that the user will be convinced that the solution is sound. This is similar to the proof of a theorem: if one executes all its steps, we will arrive at the result.

The meta-trace makes it possible to explain how a solution has been found

However, such an explanation is often insufficient, as we also want to know how someone could find this solution. We are often sure that the solution is correct, but a feeling of dissatisfaction remains: how could we find such a beautiful resolution? We are not only interested in finding a solution, we want to acquire an efficient problem solving method, so that later on we could solve analogous problems by ourselves. Many students experience this feeling in mathematics, they do not see how they could find the perfect solutions given by their teacher. Some of them come to think that only special people can do it, because they have a special gift: they have a good head for maths. They do not solve problems which are within their capability, because they do not really try to, they do not see how they could do it. They are excusable; many teachers only give a proof, but they rarely explain how one could find this proof. However, the teachers are also excusable: applying a rule, finding an intermediate result are conscious processes. Thus, it is easy to know what one has done while solving a problem, we easily generate the first kind of trace. In contrast, why we decide to apply a particular rule rather than another one is often completely unconscious, so the teacher cannot indicate why he has chosen to make such a fruitful attempt; he does not know why. The consequences of this limitation on the teaching of many sciences are disastrous; we cannot give the students the information that they need for improving their problem solving methods. We can only train them, and we ask them to solve many problems, hoping that this will help them to build an efficient method. Some succeed, while others fail, often because they do not even try to solve these problems and they believe that they will forever be out of their reach.

As in mathematics, in every domain, when we are not in a stereotyped situation, we are working in two steps: first, we select candidates, then we compare their advantages and their drawbacks, so that we can choose the most interesting one. The first step is often unconscious, while the second is conscious, thus we cannot explain why we have considered a candidate. This does not happen only in scientific domains, such as mathematics, but also for our everyday decisions. For instance, if I have to go to the airport for the first time, a huge number of actions are possible, such as to stand up on a table, or to sweep the living room. Naturally, I never consider such actions; an unconscious mechanism selects a few conceivable actions: but I do not know why I considered taking a taxi or the subway, while I do not even consider other reasonable actions such as to go on foot or to hitch-hike. The unconsciousness of the first step is dreadful, because we often do not make the best choice only because we do not even consider it; as we do not know the reason for this leaving-out, we cannot change our way of thinking so that the useful decisions will be certainly considered in the future.

An artificial being may work differently. If it is not very clever, and if it has a repertoire of rules, it will systematically try to execute them in each new situation. If so, it is easy to explain how it found a successful attempt: "I tried everything, it failed until I came to this attempt which succeeded". This can work sometimes with very fast computers, but there may be so many possible attempts that the method is not feasible; besides, that is not very useful for someone who wants to learn how to solve problems.

CAIA does not work in that way – it carefully chooses which rules it will execute. To do this, it builds a method, that is, it associates several conditions with each rule, and also a trigger, which prompts it to consider the rule. It considers a rule only if it has been triggered, and if all its associated conditions are true. We have seen the rule X+0 is rewritten as X, but CAIA will not consider it each time it tries to simplify an expression, which would not be a waste of time if it had to simplify the equation A*1+B=3*C. It has a trigger, which is the presence of zero, so the rule will only be considered when it finds a zero in an expression. Thus, the first step for choosing a rule is no longer unconscious; it knows why it has considered a particular rule: it has been triggered, and a set of conditions was true. Moreover, both the trigger and the conditions are known; thus, it can put them in a second trace. I am calling it the meta-trace, since it contains the reasons for the actions. For instance, it indicates that the solver has chosen to apply rule R4 to constraint C152 because this constraint is an equality with only two variables, and these variables appear in many other constraints. In contrast, the first trace only indicates what happened: applying rule R4 to constraint C152 generated the new constraint C314. This is possible because the triggers and the conditions associated with each rule are given in a declarative form, so CAIA can manipulate them consciously. For a human, this kind of information has been proceduralized, it is implicit, and only available when we are using it: we can observe it neither when we are using it, nor when we are searching for it in our memory. With this meta-trace, CAIA can explain why it has thought to execute the rules which appear in the first trace, and I call this explanation a meta-explanation, that is an explanation of the reasons for the choices that has been made. We will see later how meta-explanations are generated and used, but here we are only interested on the acquisition of the information that enables CAIA to create the meta-trace.

The interest of explaining "Why?" and "How?"

An explanation indicates why a solution is correct. It is a powerful tool for learning: when one has understood the steps leading to a solution, one can try to improve one's future performances in favoring the steps that were useful. The explanation can be built from the trace. The meta-explanation indicates how a solution has been found. This is also useful, but in another way than the explanation. It shows that some useful attempts were considered too late, or that some actions were not taken because they were not even considered, although they would have been very effective. With an explanation, one can see productive rules, which one can add to one's tool box. With the meta-explanation, one sees how one can build a method for adequately using this tool box. As humans cannot generate a meta-trace, they cannot produce the meta-explanations that would enable them to create a method for efficiently building the shortlist of possible decisions that will be examined thoroughly. Our impossibility of generating a meta-trace is a severe handicap for human beings, but it is also a handicap for artificial beings: although they can create a meta-trace, we have no human model for indicating how to use wholly this potential advantage.

Contrary to artificial beings, we have serious difficulties in giving meta-explanations because our meta-trace is almost completely unconscious. How many times have we said: "Alas, if I had only thought of it!" The difficulty is that, not only have we not thought of it, but we do not even know why we did not think of it. We frequently fail to do something, not because we reject it for its drawbacks, but because we do not consider it at all. We have already spoken of de Groot's experiments with chess players: in a particular chess position, all the grand masters had considered the best two moves (even when they had only considered three moves), while many good players had not even considered the best move (even when they had considered more than ten moves). One of the most important factors of the strength of a grand master seems to come from its ability to consider the best moves, and not of its ability to develop huge trees: the size of the tree did not seem related to the strength of the player. Some time ago, a chess champion had replied to an admirer who asked him how many moves he was considering: "Only one, but the best!" This is a sally, though there is a large part of truth in it, as the choice of the moves considered by a chess player is crucial; as this is an almost completely unconscious process, it is difficult to improve it or to teach it.

5.2. Auto-observation after the completion of a task

We will now examine another kind of auto-observation, which is not made just after an event, but some time later. The goal is no longer to execute a task as efficiently as possible, but to understand the reasons for the mistakes that have been made, for the clumsiness of some actions, and also for the unexpected successes. Thus, one will avoid the failures being made again, and one will favor the repetition of one's accomplishments. This work is done on the trace and the meta-trace, which were created during the execution of the task. The trace indicates the attempts that were useful, and also those that fail. It enables an intelligent being to check a solution, and leads other people to accept it. In contrast, the meta-trace indicates why an attempt was considered. Thus, one can see why an attempt has been made too late, and why some other attempts has not even been considered. With this knowledge, one can learn by improving how the rules are used: one modifies the triggers that indicate that one will look at them, and the conditions that finally decide whether they will be considered.

5.2.1. *Creation of an explanation*

An explanation is essential for a user because it enables him to verify that a resolution is sound, and so to accept the result. It is also crucial for an intelligent being: one can see among the many attempts those that were used for finding the solution. Thus, he can gather the elements that can favor the productive attempts, and avoid the fruitless ones. CAIA builds and uses an explanation for every problem that it solves, the trace is the essential element for doing so.

While a problem is solved, many intermediate results are found; some are absolutely necessary for further steps that lead to the solution, while others are a waste of time. Naturally, one cannot always know which ones are necessary when they have just been created, this can only be found when the resolution is completed. Many intermediate results seemed promising when they were found, however it happened that they were never used later in a step leading to the solution. For instance, carrying on with the example where we want to go to the airport, taking our car seemed a good idea, and several steps were effectively made: going to the garage, opening the door, entering the car, etc. Unfortunately, the car had broken down, and we could not start it. We have not to use the preceding steps for explaining how we went to the airport with the subway, they were not useful for attaining our goal. They are only needed for the meta-explanation, which will show why we chose to use the subway. This is why an explanation cannot be generated during the resolution; one has to wait for its end.

However, some intermediate results are certainly useful, when they bring an important contribution to the final result; the explanation is built from these results. Evidently, this occurs when a rule finds a solution, or states that there is a contradiction. Either this solves completely the problem, or it closes one of the cases that have been considered. The creation of the explanation will start from these events: with the trace, one knows which rule has led to this result, for instance one has found the contradiction because one has found the constraint X*X=-4, and a rule indicates that a square cannot be negative. Thus, this step is added to the explanation, and one now considers that finding the constraint X*X=-4 was also an essential step. One resumes the same method with this constraint, and one is looking in the trace for the rule which proved it, for instance it was the substitution of the value of the variable Y by 2 in the equation X*X=Y-6. Doing this substitution is another step to keep, and one now has two important events to explain: how did one establish the equation X*X=Y-6 and how did one find that the value of Y was 2? For both results, one resumes the method, and one finally stops when one arrives at the initial data of the problem: there is no need to justify them. In that way, one has generated an explanation, one has only to read it in the order opposite the order used for generating it, starting this time from the initial data. All the intermediate results which are not in this explanation were completely useless, although some of them looked very promising. Naturally, the explanation also includes the logical links between its various elements, for instance that Y=2 and X*X=Y-6 lead to X*X=-4, using the substitution rule. With such an explanation, everyone who agrees with the rules that are used will also agree with its results.

Explaining how a result has been found

The explanation generated in that way indicates the succession of actions necessary to justify the resolution, but that does not indicate why these actions were chosen. This will be given by the meta-explanation which indicates how a result has been found. For instance, because a dog had not barked, Sherlock Holmes had inferred that the dog knew the murderer. Everybody knew that the dog had not barked, but only Sherlock Holmes had deduced an important clue from this fact. Once the remark had been made, everyone agreed with him, but why was he the only one that found it? It is not enough to know a fact, which appears in the explanation, it is also necessary to know how to use it, which appears in the meta-explanation.

In order to generate a meta-explanation, CAIA uses its knowledge of the useful and the useless actions, which have been defined in this way by the explanation. For its part, the meta-trace helps it to find the conditions that triggered the idea to consider these actions. Let us give a more concrete example, which was not made by CAIA: a patient has severe headaches, nausea and dizziness, and a mild equilibrium disorder. He consults ten physicians, who find nothing, and conclude that it is due to a psychological problem. Finally, the 11th physician does not act as his colleagues, he examines the bottom of the ear, finds a bit of earplug, removes it, and the patient is definitively cured. The explanation is simple, the remains of an earplug caused an infection, which caused all the symptoms. However, we are also interested in the meta-explanation: why the first ten physicians did not examine the bottom of the auditory canal, and why the last one did. As such processes are mainly unconscious, we have no meta-trace indicating why some rules were not considered, we can only make an educated guess: the last physician had a rule that suggested examining the auditory canal when a patient had an equilibrium disorder. While the explanation indicates what happened, the meta-explication indicates how the solution was found.

In this section, we have seen that we have a mechanism for choosing the elements which will be put in a shortlist of candidates; unfortunately, as this mechanism is usually unconscious, we cannot create a meta-trace which includes the reasons for the selection or the elimination of a possible action. If the best action has not been put in this shortlist, it will certainly not be chosen later, and we cannot know why we have omitted it; moreover, an expert cannot teach his excellent method for selecting the best candidates. After this first selection, a conscious mechanism leads to the determination of the best element among those of the shortlist. As this is conscious, we can generate a trace that can be used to produce an explanation justifying the solution. An artificial being does not have these constraints: when its knowledge is given in a declarative form, it can create a meta-trace which can be used for generating a meta-explanation. It can therefore know why the elements of the shortlist have been chosen. For the second step, where it determines its final decision, an artificial being creates a trace, and generates an explanation in the same way as a human being does. Thus, this is mainly to the existence of the meta-trace that artificial beings have a huge potential of progression in comparison with us, and it is all the more important because the choice of the elements of the shortlist is crucial for the performances of a problem solver.

5.2.2. *Using an explanation*

An explanation gives a user a complete justification of the result. However, it is not very useful when this result has been found with a systematic combinatorial search: one has considered all the possibilities, and one has kept those that satisfied all the constraints of the problem. In this situation, one can only say: "I have considered 2,795,345,561 situations, and I have verified that only one was a solution." This is not satisfactory because, although we can check that the given solution is correct, we cannot check whether there is no other solution. We assume that the computer has not made a mistake, and that there is no bug in the program, which is very questionable. Most chess programs give this kind of explanation, they develop a huge tree, and assert that, if one plays such a move, one wins a Knight; as they have considered millions of moves, we trust them, but we cannot always understand why this move is so good. Chess grandmaster John Nunn has studied many chess endgames solved by the systematic generation of all the winning positions, for instance one knows all the winning positions where there is a King and two Bishops against a King and a Knight. In any such winning situation, one also knows the fastest winning move. Nunn succeeded to explain the reasons for playing these moves for many end games. However, for the endgame King, Queen and Pawn against King and Queen, although there are only five pieces on the board, he recognized that he was unable to explain why one must play the moves found by the computer. As he said: "I will leave it to a future author to deal with." It is not always possible to find solutions that are explainable because they consider too many possibilities, in some cases intelligence cannot reduce the size of the search tree enough to be understandable by us. Anyway, CAIA first tries to generate a solution with few cases to consider, so that we can understand them: it manipulates the formulation of the problem, so that the combinatorial search is made only when nothing else is possible. Thus, for most of its solutions, it produces explanations that we can verify and understand.

Let us take an example showing various explanations for various ways of solving a problem. This problem, also presented in Appendix 1, is to find all the possible values for the integers A and B that satisfy the constraint $4*A + 3*B^2 = 34$, B being greater than -1,000,000,000 and lower than +1,000,000,000. It is easy to write a combinatorial program which will consider the 2,000,000,001 possible values of B, for each of these values it will compute $34-3*B^2$, and it has found a solution if this number is a multiple of 4, that will give the value of A. For instance, if B=7, the constraint becomes 4*A=34-3*49, that is -113, which is not divisible by 4, so this is not a solution. If we find a pair of values for A and B satisfying the constraint, the program prints the result. Thus, once this program has been written, we run it on a computer, and after some time, it stops without finding any solution. The explanation is: every possible solution has been tested, and none led to a solution. Nevertheless, are we sure that there was not a mistake somewhere? We cannot ourselves perform billions of computations. CAIA gives a completely different explanation for its resolution: $3*B^2$ is even because it is the difference of two even numbers: 34 and 4*A. As 3 is odd, B^2 is even, then B is even. As B is even, B^2 is a multiple of 4, 4*A is also a multiple of 4, their sum is a multiple of 4, which cannot be equal to 34, which is not a multiple of 4. This is a contradiction, so there is no solution. Everybody can check this solution, which is very simple since it uses twice the same rule: if all the elements, but one, of the equality of two sums are divisible by an integer N, then the last element is also divisible by N.

An explanation can be used for learning

An explanation is precious for a user, who accepts the solution, but it is also useful for CAIA itself, because this enables it to learn. Analyzing the explanations, CAIA counts how many times each rule was useful, and how many times it was useless. Then it will favor the use of a rule if the first number is high, and it will consider it less often if the second number is high. It also uses the explanations for improving the directions for use associated with each rule; this considerably decreases the time necessary to find a solution. The kind of learning that it utilizes here, called Explanation-Based Learning (EBL), is powerful because one can learn from a single example, instead of from statistics on thousands of examples. In this kind of learning, the first step is to build an explanation, so that one understands what happened. Then, one tries to generalize the situation where a fortunate event occurred in such a way that all the conditions for the first success still stand: in this way, one will try to avoid an unsuitable action. To give a concrete example, let us assume that we barely escape from being hurt by a car when crossing a street. We could merely decide to be more cautious the next time that we cross this street. However, it is better to also understand why we were almost injured. We therefore analyze the event, and we discover that the reason was that the color of the traffic light was different for the two directions (that often happens in Paris): it was green for one direction, and red for the other, where the cars had stopped. Thus, as the cars were motionless, we began to cross the street, and we were almost knocked over by a car coming from the other direction, which still had the right to drive. After this analysis, we have understood the danger of this situation, and we decide to watch the traffic lights for both directions before crossing a street. From the understanding of the reasons for a single incident, we have built a method which can be used in other places than where it happened.

To summarize, explanations give crucial information for creating and improving the directions for use of the rules which control one's behavior. An artificial being will be more efficient when it considers first the rules that will likely succeed. Finding which rules are successful can be made by an inspection of the explanations. This examination can be made more cleverly than only counting the successes and the failures: one can classify the contexts where a particular rule succeeds or fails. For instance, we could see in the explanations that a particular rule often gives a useful result when it is applied to equations with at most three variables, and that it always fails when applied to equations with more than five variables. So, one can improve the way this rule will be considered later on.

Using a meta-explanation

CAIA also gives meta-explanations, that is, it associates with each useful action the reasons why it was considered. This is interesting for the solver, once one has understood these reasons, one can use them for improving the way one chooses which actions one will think over. One can even do it consciously now that one knows some methods for selecting the actions to consider. CAIA can also indicate some blunders that it has made. For instance, it indicates that it tried to take advantage of the simplicity of the equation $X=2*Y+3$ in substituting X by $2*Y+3$ in the equation $X*Y+3*X^2=5*Z*X^3$ because the first equation is linear while the second equation has only three variables, which include the two variables, X and Y, of the first equation. Indeed, one of the triggers for considering the substitution rule is the discovery of a linear equation with only two variables, and the choice of the second equation, where the substitution was made, only required the two preceding conditions. I do not write the result of this substitution which is a constraint that is too complicated to be useful, thus this step was useless. CAIA is not still clever enough to use this information itself, but I could use it: I did not change the trigger, but I restricted the choice of the second equations by adding conditions on the number of its terms and on its degree: for instance the second equation must be linear, and contain at most two variables. When using a meta-explanation, a human can learn to improve his method for choosing its attempts, and both human and artificial beings can modify the directions for use of their rules so that less unsuccessful attempts will be made.

Explaining why a result has not been found

It is even possible to give "Why not?" meta-explanations. A Why not? meta-explanation does not justify a result, but indicates why some results have not been obtained. One cannot always give such explanations, because there may be too many ways to explain the absence of a result. It is impossible to answer to a question such as: "Why am I not the King of England?" there are a huge number of possible answers from "because I was not the son of King George VI" to "because I am French". One can answer to such questions only when there is a near miss, that is the fact in the question is almost true. The question "Why did not I win this chess game?" is important for a chess player, and it can be answered because it was quite possible that he won if both players had the same strength. Thus he replays the moves of the game, until he finds, for the first time, another move which could improve the situation.

CAIA works in the same way as this chess player when it has solved the same problem with two different methods, and one resolution is more elegant than the other. We have already seen that a method is the set of triggers, of conditions, and of priorities associated with each rule; it determines whether and when a rule will be considered.

I will show the interest of this kind of meta-explanation with the following example, which really occurred. CAIA had found, using a method that it had created, that the value of the variable X is 7; in contrast, using the method that I had given, it had not found this result. Therefore, it wondered: "Why did not I see that the value of X was 7 when I used the second method?" In order to answer this question, it searched in the first resolution for the action that established X=7. From that point, comparing both traces and meta-traces, it looked for the last result that leads to X=7 in the first resolution, and which is also present in the second resolution. Up to now, the wrong resolution was correct, since it found the same results as the right one. The missing action is the action that was applied to this result in the good resolution, and which later led to proving X=7: it was not executed in the poor resolution, since its consequences do not appear in the trace. Either this action was not triggered, or at least one of its conditions was not true. In this example, CAIA discovered that it had not found X=7 although it had established the equations 2X+3Y=23 and X+Y=10. The missing action, made in the good resolution, is that it did not take the value of Y in the second equation, then substitute it in the first one, which gives the value of X. When CAIA looked at the conditions associated with the substitution rule, it unveiled the culprit: I had included, in the conditions for agreeing to make a substitution in a constraint, that this constraint must have at least three variables. As the second equation had only two variables, it was not selected, so the value 7 for X was not found. Naturally, I put right the condition, and now my method also accepts making a substitution in an equation with only two variables. In this situation, using its ability to create Why not? meta-explanations, CAIA helped me, showing a blunder which I had made when I defined my method.

The explanations indicate the sequence of rules that were necessary to solve a problem, thus they justify the solution. The meta-explanations indicate why these rules were considered, and make it possible to see that they have been utilized clumsily. The first kind of explanation depends on the trace, while the second one also depends on the meta-trace; both can be generated during the resolution, because CAIA is conscious of its actions, and of the reason why it considered them. These traces enable an artificial being to improve its knowledge, and the way to use it.

More on the subject

John Nunn has written three books where he explains the results found by a combinatorial program, which generated all the winning positions when there are at most five pieces on a chess board: *Secrets or Pawnless Endings*, *Secrets of Minor-Piece Endings*, and *Secrets of Rook Endings*.

I have realized the first program using the EBL method in a program learning to play chess. For instance, from a single example where a piece was captured because it was pinned, it learnt the interest of using a pinned piece, and it could use it in other situations. This system is described in J. Pitrat, *A Program for Learning to Play Chess*.

5.2.3. *Finding anomalies*

We have seen that some anomalies were discovered during the execution of a task, either by the operating system, such as an attempt to divide a number by zero, or by CAIA itself, as an excessive memory consumption. When this occurred, the task was stopped, after storing all the available information; this will be used later on for a better understanding of the reasons for this anomaly. However, we will now see that it can find other anomalies when the task is completed.

CAIA performs many resolutions for each problem

An obvious anomaly appears when CAIA obtains different results when it uses different methods to solve the same problem. Indeed, a problem can be solved more than 30 times, each time in a different condition. In particular, we have seen that I had given CAIA a method, and it has itself created two other methods. These methods modify the formulation of the problem, and they also use tricks, which can eliminate many possibilities, so that they usually find solutions which can be explained to a user, and which are as elegant as the solutions found by human experts. Thus, CAIA will successively solve a problem with each of these three methods, and compare their performances. Moreover, it can also write a combinatorial program, which considers many possibilities, and so gives solutions that cannot be understood and verified.

However, the main source for different resolutions is the experiments that CAIA makes to see whether it would not be possible to improve its problem solving method. We have already seen that CAIA may solve a problem with different personalities. In some cases it wants to find a solution quickly while, in other cases, it is very stubborn because it wants to find a very elegant solution, it is no longer afraid to try dangerous rules which often waste a lot of time, but sometimes succeed unexpectedly. Many experiments are also made when CAIA wants to examine the way it is doing proofs by case. For instance, at some point in the resolution, the variable X may have two values, 3 or 5, while the value of variable Y is 1 or 2. When there is no other possibility than choosing a variable, and looking what happens for each of its possible values, it considers first what happens if X=3, then one with X=5. However, it will also try Y=1, then Y=2. In some situations, the differences between the simplicity of the solutions depending on choosing X rather than Y is tremendous. Thus, one can find a surprising difference, which can be used to learn to improve one's ways for choosing the variable. CAIA can also decide to solve a problem again, but this time it inhibits a particular rule: it wants to see whether this rule is useless although one can often execute it. All these experiments can lead to a huge number of different resolutions for the same problem, this is the routine of a scientist's work, who/which makes many attempts, analyzes them, and detects surprises. These studies can lead to a better understanding of the best way for completing tasks in his/its domain.

Anomalies discovered by the comparison of several resolutions

Thus, CAIA carefully analyzes the results of all these resolutions, and it can find many anomalies. The most evident anomaly occurs when the number of solutions is different for two resolutions of the same problem: if one method has found 5 solutions, and another one 6 solutions, there is at least a serious error: either the first one has missed a solution, or the second one has found an erroneous solution. We have already examined what CAIA is doing when that happens. However, there are many other anomalies that it can detect: each time it sees an important difference between two resolutions, it notes it, and reports it to the leader of the bootstrap. For instance, we mentioned in the last paragraph the importance of the manner of choosing the variable used in a proof by case. If the tree generated after choosing variable X has 2,000 leaves, while after choosing variable Y it has only 15 leaves, if CAIA had chosen X rather than Y, this is an abnormally poor choice, and this will be signaled. Naturally, if CAIA had preferred Y to X, this would no longer be an anomaly, but an excellent decision.

As CAIA can be conscious of the reasons of all its actions, it knows its belief that a particular decision is good, thus it can make a sharper analysis of an anomaly. To explain this, let us consider again the preceding example. Preferring X with 2,000 leaves to Y with only 15 leaves is a blunder, but it could be worse: if it had hesitated between both choices, it is not so serious that if it had considered *a priori* that choosing X was much better than choosing Y. Considering what is bad as much better than what is good is much worse than judging them equivalent. Thus, it also uses its faith in the quality of a choice for evaluating the seriousness of an anomaly.

CAIA also finds anomalies when it compares the results of two methods on the same problem. We have said that it can use a clever method giving solutions similar to human solutions, and also write a combinatorial program, which give the kind of solution that the computers usually give, that is fast and incomprehensible. Usually, this clever method is slower but, if it is much faster, this is an anomaly which is remarked: for some problems, it happened that the clever method solved the problem in 15 minutes, while the combinatorial program had only developed a small part of the tree after 10 hours. CAIA notices and reports such facts; this is an example of a fortunate anomaly, where there are surprises which are not due to a blunder.

An interesting anomaly: a record

A particularly interesting kind of anomaly is a record. Humans are also interested in records, there are even books specialized in this subject. Many records found by CAIA come from its capacity to create new problems, for instance there are records for the difficulty of a problem. This can be the difficulty to solve it, which can be evaluated by the smallest time used by all the methods that have been tried. This may also be the difficulty to understand a solution, evaluated from the smallest number of leaves of the tree developed for finding all its solutions when it used a clever method. The reverse record, finding the problem solved with the smallest tree, has no interest: many difficult problems are directly solved by CAIA, without any backtrack. There are also records for families of problems, for instance an easy problem in a family of difficult problems, or a difficult problem in a family of easy problems. For instance, almost all its Sudoku problems were directly solved, a few needed a tree with two leaves. If it ever finds a Sudoku problem that it cannot solve without developing a tree with at least three leaves, it would appreciate this record. The adviser examines if several records have been beaten since its last examination, and it takes that into account in its evaluation of the performances of CAIA during the period that it considers. This helps it to find whether it is always in a productive research phase.

All in all, CAIA can spot a large variety of anomalies when it examines the results of different resolutions of the same problem, and also when it compares the performances of various methods for families of problems. It does not yet know how to benefit from these analyses, but it points them out to its human leader of the bootstrap; this is a very effective help, because we have not the time to perform and analyze so many experiments. In the collaboration between a human scientist and an artificial one, this is perhaps the point which is the most important. To be surprised is essential because, when it has been understood, one has ideas for improving one's knowledge on the domain. The experiments carried out by CAIA effectively lead it to detect many anomalies, which were pleasant or unpleasant surprises, but which were the source of fruitful thoughts for me.

Chapter 6

How to Observe Oneself

We have shown why it is important for an artificial being to observe itself; we now have to show how that can be implemented. This chapter is more technical: it is not sufficient to say that an artificial being can have many extraordinary activities related to consciousness. I also have to indicate how that has been done. I will try not to give too many details, but to describe only the main principles, which are the core of these realizations. However, for the interested reader, I give more details on the implementation in the appendices.

To perform all the observations that we have seen before, an artificial being first determines when it will observe its behavior. It risks missing an important point if it observes itself too rarely, or wastes too much computer time, and also gathers too much useless information, if it does it too frequently. Secondly, when it stops to observe itself, it has to know its last actions, and everything that may be useful about its present state. Thirdly, the information that it has obtained is examined for storage of some useful bits, and also for possibly modifying its way of fulfilling the current task. CAIA is not the only being interested in observing the behavior of a computer system; a human programmer needs the same kind of information when he is debugging a program. Therefore, the method used by an artificial being for observing itself has already long been widely used for debugging programs. The only difference is that it is the artificial being, rather than the programmer, which uses the information gathered in this way.

We begin with a description of three techniques which make it possible to stop when a task is running, and to search for any potentially useful information which could be used immediately, or later on when the execution has been completed. It is always better to examine this information as soon as possible, so that one can act and avoid carrying on in a hopeless direction; in that way, one can modify what is wrong immediately after spotting an abnormal situation. Moreover, the observations are better collected when they are used as soon as they are available, because one can dynamically modify what one wants to observe, taking into account the latest facts. In an analogous way, when a crime has been committed in a village, the policemen try to check the alibi of a tramp who was loitering nearby, but they will switch their attention to the family and friends of the victim, as soon as they learn that the dog did not bark.

Observing during a task or after its completion

Auto-observation can be done while a task is normally running, the artificial being only looks from time to time at what it is doing, but *a priori* it does not expect trouble, it is only cautious. However, it may occur that it observes itself more completely, because this is not a normal task: it has seen during a previous execution a serious anomaly, and it reruns this task to explain why this happened. Finally, there may be an unexpected problem, found either by the operating system or by the monitor agent. In this situation, it gathers information, and decides whether to stop, or to complete its task after modifying its future actions. Thus, the method that it chooses for observing itself is adapted to each of the three following possibilities: routine control, searching for the cause of a previous problem, or understanding a completely unexpected situation.

Some analyses can be made only when a task has been completed because it is impossible to evaluate the interest of an action before one has seen all its consequences. The information that could be inferred from the observations entirely depends on the quality of the trace that has been stored: one cannot understand the reason for an anomaly if one has not bothered to keep a fact which appears to be essential when the difficulty becomes evident.

When an artificial being begins a self-observing phase, it knows its internal state: without this, it could never realize whether the current task is progressing satisfactorily. It has also access to its knowledge so that it can examine it, and see if some bits are incomplete, erroneous or missing.

A fairy tale

In some tales, a fairy uses her magic wand to freeze everybody; people stay exactly in the position where they were when she used her wand, and they stay so until she uses it again. Then everybody carries on with his interrupted task without even noticing that it has been stopped for a while. During this period some happy few have not been frozen, and they can examine the frozen people. With the three methods that we describe in the next sections, one can be in the same situation as these happy few, one can examine everything in one's present state, and memorize it. If necessary, one can also make some changes if one sees something wrong in the present situation. Moreover, one is not only one of the only happy few who is not frozen, but one is also the fairy: one decides when it is necessary to freeze everything, and one also decides when one restarts the current process because one has observed all that one wanted to know. As in the tale, when the process restarts, everything happens as if there had been no interruption, except naturally if one has modified the situation by correcting an erroneous object.

We will successively examine these points, beginning with the methods which allow an artificial being to determine when it will choose to observe what it is doing. We will first describe three methods which entirely depend on CAIA itself: interpreting what it is doing, adding examination points, and using interruptions by the clock. A fourth method depends on the operating system, which signals an anomaly. Then, we will see how an artificial being can observe itself when it has decided to do so, and particularly its present state and its knowledge. We will show that the existence of a society of mind with several agents considerably simplifies the implementation of this reflexive behavior. We do not forget the focusing of attention, which allows beings to choose what they are observing, so that they do not waste this capacity on minor tasks. In the last section, we return to a point considered in the foreword: the use by artificial beings of a "human" word, here related to consciousness. Finally, the appendices have been written for readers familiar with computers, so that they can understand how these possibilities have been implemented in CAIA.

6.1. Interpreting

A computer executes the instructions of a program, written in a language which it can decode. One can give it such programs directly, but that is not convenient because one would order its future actions in much too detailed a form. Thus, many languages have been devised so that it is easier to communicate with a computer: some examples of these languages are Fortran, Algol, C. The programmer writes SPACE=SPEED*TIME instead of writing a sequence of instructions ordering to take the content of memory 3564, to multiply it by the content of memory 6587, and to put the result in memory 2964. We no longer use the number of a storage cell, the variables may have mnemonic names, and the orders are written in a form similar to a mathematical language. The preceding order indicates that a variable, called SPACE, is the product of two variables, SPEED and TIME. The values of the last two variables must have been defined before, and this instruction defines the value of variable SPACE.

The compiler

As a computer is not built to execute such orders, another program, associated with the particular language and with the computer which is used, will ensure that programs written in this language will be correctly executed by our computer. Often, this program, then called a compiler, translates the orders given in the advanced language into a sequence of instructions written in the computer language. For instance, a C compiler for computer type XXX will translate any program written in the C language into a program in the machine language of XXX computers, so it can be executed by any XXX computer. This translation is done once, and we do not give the compiler the data of a particular problem, the result is independent of these data. Then, one keeps the translated program, and one can use it as often as necessary with various data. The best compilers are doubly optimized; the translation is made very quickly, and the generated programs in the computer language also run very fast with the data that they receive for each execution.

The interpreter

However, there is another possibility than compilation: one no longer translates at the beginning the program written in an advanced language, one keeps it in that form. To utilize it, one uses another kind of program, called an interpreter. An interpreter for programs written in a particular advanced language, let us call it L, receives the program written in L and its data; it does not make a preliminary translation, and it immediately begins by executing the first instruction of the program. The interpreter is constantly executing a loop, which includes three steps: it decodes the current instruction, it simulates its execution, and it determines the following instruction. In a compiler, one decode an instruction in the advanced language only once, so that one can translate it, and this translation may be used billions of times. With an interpreter, an instruction is decoded each time it is used. Let us again consider the instruction SPACE:=SPEED*TIME, and assume that the value of SPEED and TIME are 3 m/s, and 10 s. The interpreter will decode this order, and see that it requires making the product of two variables, SPEED and TIME. To execute this order, it will search for their value in its memory, find 3 and 10, perform the product, find 30 and store it as the new value of variable SPACE. The third step is trivial in this situation; the next instruction will be the following one in the program. From this point on, it can use the new value of SPACE, until it may be modified again. The interpreter acts exactly as a human being who can also simulate the execution of this instruction.

Comparison between compilation and interpretation

Let us compare a compiler and an interpreter when both are applied to a program P written in an advanced language. The compiler translates P into a program Π written in a computer language. The execution of Π, applied to the data D, gives exactly the same result as the interpreter when it simultaneously receives as data P and D. With the interpretation, it is no longer necessary to spend some time in the translation P into Π, but it is much slower since one has to execute several hundreds instructions of the interpreter for each instruction of the original program, instead of 3 or 4 instructions of Π. If P will be executed many times, the time necessary for the compilation becomes negligible, and the interpretation will use much more computer time. We can give a good estimation of the slowing down due to the use of an interpreter: it is about 60 times slower than when one uses a compiled program such as Π. Thus, the seconds become minutes, and the minutes hours; the interpretation appears as a waste of time, since it repeatedly decodes the same instructions. Why is it useful to do a sequence of instructions billions of times that could be done only once? This is often true, and that is why compilers are widely used.

However, on some occasions, an interpreter has a huge advantage over a compiler because it regularly receives the control of the execution, and it is independent from the executed program P. Thus, when it executes an order, it can examine not only the current order, but also the value of all the variables, and the progress of the program P which it is interpreting. It can also see that it is dividing a number by zero, note the events that seem important, gather statistical data; moreover, it can dynamically modify the program P, so that it will be more efficient and will contain fewer mistakes in the future. It can also give control to the user if some conditions are met, so that one can examine all the characteristics of the interesting situations, selected by the interpreter, and see the reasons for an anomaly. At present, I am always the user of the interpreter, but in the future I hope to give this possibility to the monitor agent of CAIA. An interpreter is a very useful tool for debugging and was one of the first programs ever given to a computer, even at the time when computers still had no operating system, and no compiler.

Accelerating the interpretation

We have seen that the main drawback of interpretation is its slowness, but it is possible to remedy this defect. Indeed, one often knows which subroutines will be watched. Most of them can be considered safe, or at least it is unlikely that they produced the anomaly that we want to understand. Thus, the process is running normally, as it was produced by the compiler, until one comes to the suspected subroutine. Then, instead of executing the compiled version, one calls the interpreter, and asks it to execute the original version with the current values of the variables. Once the interpretation is achieved, and one has observed all the dubious steps, one resumes the normal execution of the compiled program, until one arrives at another meeting point. One can even give conditions for beginning the interpretative mode, for instance one can order to interpret a subroutine evaluating a patient's health only if he is a man, more than 60 years old, and at least 1.80 meters in height.

With this improvement, an execution including a limited interpretation is almost as fast as a normal execution. It is only slowed when the interpreter asks the advice of a human observer, who needs a lot of time to make his observations. There is an interpreter for the language used for giving CAIA its knowledge; this is crucial for helping me to debug the system. I widely use this interpreter, but CAIA does not use it yet, although it gives it a huge potential for observing itself. The reason for this temporary underutilization is that the methods used in the two following sections give it enough possibilities in that domain for the present time.

Advantages of the interpretation

All the same, I am so amazed by the possibilities which are offered by the interpretation that I will certainly some day give CAIA the capacity for observing itself in that way. An artificial being can memorize any interesting event with its context, observe the state of the system, including its knowledge, modify the value of variables so that the future behavior of the system will be improved. It can even modify its own programs, including the subroutine that is currently interpreted; these modifications will be immediately taken into account. These possibilities are extraordinary; the control is regularly given to a consciousness that can observe everything, analyze these observations, and then modify any part of the artificial being itself. Once these observations and these modifications have been completed, it can restart the current task, which will not be disturbed by this interruption. However, one must be cautious with real-time applications, such as driving a car: in that case, it can certainly not spend too much time observing itself, if it does not want to risk an accident. However, for many human activities, such as proving theorems, playing chess, managing a firm or a country, and so on, these interruptions are not ruled out for a present computer, as its speed makes them short.

The possibility of writing a program that could interpret any program, including itself, is also important from a theoretical point of view. The results found by Alan Turing, and especially his theorems on the limitations of machines, are based on the definition of a theoretical machine, which he described in his 1936 paper. He particularly studied one of these machines, the universal machine, which is precisely an interpreter for any Turing machine. It receives as data two elements: the description of a particular Turing machine and its data. Then, it will give exactly the same result as this Turing machine applied to these data. It is the power of this universal machine that leads to the proof of these theorems.

The interpretation is a perfect way for observing the behavior of an artificial being, by its user as well as by itself. The difficulty is to know where to look, since one can see something only if one wants to see it. Moreover, it is not sufficient to observe something, it is also necessary to analyze it, and take advantage of it. This is not easy when the result is completely unexpected, and CAIA is not yet able to do this last part. However, for some time now, CAIA can selectively interpret only what is interesting, and dynamically accept modifying its center of interest, This possibility gives it an adjustable consciousness, which can observe everything, with as many details as it wants, without wasting much time. This adjustable consciousness would be very precious for human beings: nothing can be hidden, one can exactly focus on the interesting point, one can change it if desirable, one can stop to observe oneself as long as one wants, and one restarts the task as if there had been no interruption. What more could we wish?

6.2. Adding supplementary orders

Another method for auto-observation is adding supplementary orders, which can have three roles. Firstly, they can generate a trace, which will be used at the end of the execution. Secondly, they can define the values of variables, which will indicate the state of the system if it stops unexpectedly; in that way, the observer can easily know the current task and its present state of advancement. Finally, they can give control to an observer, which may be a human being as well as an artificial one. He/it can examine all the parts of the system, and possibly modify some parameters or some expertise, so that the future behavior of the system will be improved. When the observer estimates that one's task is completed, one restarts the system, which will give exactly the same result as if it has not been interrupted, except for the case where the observer decided to modify it.

Generating a trace

The easiest method is to add orders that write the value of some variables in a file. This is the usual way for generating a trace, which can include the value of important variables, some intermediate results, and the indication that one has been at a particular point in a program which is particularly watched. One can add conditions to this order, so that it will be executed only if they are true, for example if the value of the variable WEIGHT is at least 95 kg. Adding such orders do not slow down the execution significantly, but this has three drawbacks. Firstly, the trace may be too cumbersome if one has not selected the output carefully. Secondly, this trace is considered only when the execution is completed; thus, if there is a disaster, if one enters in an infinite loop, the computer will produce a huge quantity of uninteresting data rather than immediately acting to escape from this problem. Thirdly, these orders are put at predetermined places; if one takes a path that does not cross any of these places, one becomes unconscious. That may particularly happen if one loops on few instructions forever.

Another method is to create an implicit trace, where one never writes in a file: one only changes the value of some variables which have been created for this purpose. Their value is only considered when there is an anomaly: then, an observer, which examines their value, has a precise indication of what a system was doing when it was stopped because this anomaly appeared. These variables are not used at all in a normal execution, as they only facilitate the work of the observer. For CAIA, some of these variables are defined by the leader of the bootstrap, but it may itself create some of them. Let us consider the case where it receives a signal from the operating system, indicating that it has tried to make a forbidden action, for instance it wanted to execute an illegal instruction, or to write in an area which is not allowed to it. If this happens, it stores the task and the program where this event occurred. Later on, if it decides that it is important to understand this anomaly, it executes this task again, but it rewrites the program where the anomaly occurred. It defines a new variable V, and before each instruction of the old program, it adds an instruction giving a value to V, this value being different each time. For instance, it adds V:=1 before the first instruction, V:=2 before the second one, and so on. When the operating system again signals the anomaly, the observing module of CAIA looks at the value of V, and it knows exactly where the faulty instruction is: if its value is 47, it was while executing the 47th instruction of the program. It is not able to complete the debugging by itself, but it has given precious help to its human colleague, who knows exactly where the error occurred.

Writing and executing new programs immediately

I take advantage of the preceding example to mention an important capability of computers, which is not directly tied to consciousness, but which considerably increases the interest of self-observation: it is not sufficient to observe oneself – one has to use these observations as soon as possible. This capability is the dynamic loading of programs. With this mechanism, a system which has modified one of its programs, or has written a new program, can immediately use it: it is not necessary that it stops and restarts. This is what happened in the preceding example, where CAIA added to an old program many instructions giving various values to a new variable V. CAIA constantly uses this possibility: during one of its lives it writes thousands of new programs which it executes immediately. This is not a secondary activity; the total number of instructions of all the new programs which it writes during one of its lives is larger than the number of instructions of all the programs making up CAIA when it starts one of its lives. It is acting as a human scientist, who writes programs to solve problems, executes them, and analyzes their results. One situation where it is writing programs occurs when it decides to solve a problem with the combinatorial method; to do that, it creates an efficient C program which only solves this particular problem, and it immediately executes it. It also makes experiments in rewriting old programs, for instance when CAIA observes something wrong, and wants to improve its behavior. In that case, it is sometimes sufficient to modify the value of a parameter but, if the modification is considerable, one has to implement it in rewriting the questionable program. Thus, the possibility of dynamically loading programs is an essential complement to consciousness.

Monitoring the execution of a task

The generation of a trace is insufficient in many situations, because it is examined only at the end of a task; thus, it is useful to also add orders that will break the normal execution. For instance, when the solver agent of CAIA comes to such points, it stops and the control is given to the monitor agent. This agent will examine the present situation, and the progress made since the last time it was called. In this way, it is acting as the fairy that we saw at the beginning of this chapter. It will decide if the solver will continue without changing anything, or if it will change something in its problem solving method. It can even decide to stop this execution, when it seems without hope of success; this will be considered again under other conditions later on.

The introduction of regular meeting points of the solver and the monitor is easy to implement. Indeed, the resolution is made by a succession of cycles, each one beginning with the choice of a rule with its arguments, then the rule is applied, the results are analyzed, and the cycle resumes. So, every N cycles, for instance every 200 cycles, the solver calls the monitor for advice. If it has created too many new constraints, it tries to understand why. If most of them are uninteresting, it decides to eliminate the less interesting, and to be more restrictive in the future: it raises the threshold that determines whether a new constraint will be kept. If too many solutions have been found, it will perhaps ask the manager whether the execution could be stopped: when a problem has millions of solutions, such as finding the circuits of a Knight on all the squares of a chessboard, finding all of them is time consuming for an irrelevant result. Once this examination is completed, the solver resumes its resolution, perhaps with new orders such as refusing to apply a particularly useless rule. In this way, the solver will change its behavior, thus it is possible to react quickly when a difficulty appears. The monitor can also change the value of N, which defines the number of cycles between two meetings: if something seems dangerous, the monitor may decide that the next meeting will be in 50 cycles, instead of 200; it will return to the old value if the situation improves.

The adviser monitors the career of CAIA

The preceding surveys happen during the resolution of a particular problem. However, CAIA also watches to see that its career as a scientist proceeds smoothly. This is the part of the adviser agent, which is regularly looking at the progress of the research since the last time it was called. Here the meetings are not defined by a number of cycles, because one cannot use the number of tasks that have been considered. The reason is that the duration of the tasks may be very different, ranging from one second for some tasks to 20 hours for other ones; in contrast, for the solver the cycles have roughly the same duration. Thus, as the number of tasks is not significant, I preferred to take into account the time since the last meeting; for instance it plans to meet its adviser every two days. When it is called, the adviser examines what progress has been made since the last meeting: a problem, previously unsolved, has been solved, new interesting problems have been created, some records have been beaten, for instance a new problem is extraordinarily difficult, some experiments were fruitful, particularly some hopes have been unexpectedly realized, etc. If the progress is low, that means that CAIA is running out of ideas: either it stops its research, or it radically changes its methods. Instead of monitoring the resolution of a problem, the adviser monitors the life of CAIA. Unfortunately, it is not as developed as the monitor agent: it only observes the manager agent, and informs the leader of the bootstrap, but it does not as yet modify the manager. When the adviser has completed its task, the manager resumes its work as if it had not been interrupted.

CAIA cannot yet determine these meeting points by itself, they are always chosen by the leader of the bootstrap. Naturally, they are chosen so that the situation is easy to analyze at these points, one is never in the middle of rearrangement of the knowledge or of the storage. This constraint makes CAIA's work easier when it analyzes a situation.

6.3. Using timed interruptions

When we are deep in thought, we become unconscious of the passing time, we forget to remove a dish form the oven, or we miss an appointment. Fortunately, if we have set up a timer, it will give a signal which interrupts us, whatever we are doing. This recall us that an urgent task is waiting in our diary.

This possibility is also essential for a computer; for instance, it executes several processes simultaneously, and it would not be acceptable that one of them monopolizes all the time, while the others are stopped. One cannot rely on the kindness of the currently active process, even if it does not want to keep the computer for itself. For instance, it may be in an infinite loop, so it cannot stop by itself because it will never arrive at the instructions that would take the decision to wait, for leaving the place to other processes. Thus, a mechanism outside the running process has to stop it when it has exhausted its allocated time. In practice, the control is regularly given to the operating system, which is independent from the other processes; thus, it can examine what they are doing. It can decide to let the processor run the current process, or to stop it and leave some time for another process.

In order to interrupt a process whatever happens, the computers have a clock, and one can indicate to use it in a program: one writes an instruction requiring that, in N seconds, the control will be given to a particular program, which will know what it has to do when this kind of interruption occurs. It can examine the situation, take a decision, and make a new appointment for several seconds later. This addition of instructions setting an alarm is crucial for a computer; without them one could not have operating systems such as they now are. Fortunately, using the clock in that way is not only reserved to the operating system, all the processes can use it. That can be very helpful, because a process can see that it is in an infinite loop, and so decide to break this loop. As these loops have serious consequences, and as they arrive too often during the execution of programs, I will examine this difficulty in more detail.

A pitfall for computers: looping for ever

Too often, a program enters in a loop while it is in a debugging phase, and even when it has been thoroughly tested, and has seemed to behave appropriately for a long time. I will show with an example how this may easily happen. Let us assume that we want to enumerate all the numbers between 1 and an integer N which are prime numbers. Naturally, there will be a loop which will successively examine all the possible numbers, and check whether they are prime; the value of a variable, let us call it V, will be the current number. A weird mind may decide to write the program in such a way that the value of V is N at the beginning. The program checks whether V is a prime number, if so it prints this value, then it considers the preceding number, giving to V the value N-1. It resumes the same procedure with this new value for V, and carries on, considering each time the predecessor V. It will stop when the value of V becomes equal to zero. This program may give a correct result, but there are many opportunities for a mistake. Firstly, the test checking that V arrives at the value 0 must be V<=0, the program stops if the value of V is negative or equal to zero. However, a programmer could write this test as V=0, thinking that the program will stop when the value of V will be equal to 0, so it is not necessary to test the possibility of negative values. Furthermore, normally a program also verifies that it receives correct data, thus it tests at the beginning that N is positive. However, many programmers may forget to include all the necessary verifications in their program. Even if these inaccuracies have been made, the program will give the wanted results as long as the initial value for N is a positive number.

Unfortunately, the day where it receives a negative value for N, for instance -13, because the user has made a typing error, a program with the preceding inaccuracies will loop forever; it begins with V=-13, checks if it is a prime number, tests that V is not equal to zero and then it will decrease V to -14, then to -15, and so on. It will never stop because V will never be equal to 0. In this case, it is easy to discover the mistake, but it is often very difficult to find out that a program will enter in an infinite loop when some rare conditions are met.

As CAIA works for several weeks without any human intervention, we must absolutely avoid it wasting all this time looping on ten instructions. We have seen that it can easily detect large loops when they include an observation point; for instance, the monitor agent is called each time 200 rules have been executed; as it verifies that there is progress in the resolution, it will see that it stagnates. Then, it will take some measures that will stop it always doing the same thing, and it will also store that this incident occurred. In the same way, when it is interpreting a program, a well designed interpreter can see that it is always interpreting the same orders, and it can decide to stop this interpretation also keeping the information needed for later understanding what happened. However, an infinite loop may occur in other situations than the preceding ones: it repeatedly executes ten instructions in an obscure subroutine, and there is no observation point among those instructions. Thus, any system in this situation, including CAIA, becomes completely unconscious, and it will stay so forever if there is no possibility of an exterior intervention.

Zeus detects infinite loops

This exterior intervention can be made by a human user, but the system would no longer be autonomous, it would depend on a human observer. Fortunately, a clock interruption enables a system to call a particular program independently of what happens in the main program, even if it is pursuing an infinite loop. In that way, at the beginning of a new life, CAIA indicates that, in ten seconds, one of its agents, Zeus, will be called whatever happens elsewhere. Thus, after ten seconds, Zeus is activated, and it examines whether CAIA has progressed. CAIA has been designed so that Zeus will be able to answer this question. Using a method seen in the preceding section, several variables which describe its present activity have been defined. This is a type of supplementary order, which is only useful for giving information on the present state of the system in such a way that it will be easy to find and to use it. In particular, some of them store the current task, the number of rules already executed, the maximum time allocated for this task, etc. Another variable, let us call it V, is incremented each time CAIA met an observation point. Thus, if the value of V has increased since the last call of Zeus, it knows for sure that it is not in a loop without any point of observation; if it loops, it regularly comes in a conscious state. If so, other methods are in charge for discovering and solving this difficulty. In contrast, if the value of V is the same after 10 seconds, it is more than likely that it is in a loop: there are many points of observation in CAIA, so it is almost certain that it cannot be unconscious for such a large time, in 10 seconds a computer makes a lot of work. Anyway, to be sure that it will not interrupt a perfectly normal process, it notes that there is a loop presumption, it still restarts CAIA for ten more seconds. If, at the following call, Zeus observes that the value of the variable V is always the same, it is almost completely sure that CAIA is in an infinite loop. Thus, it stores all the information on its present state, mainly the task, the advance in the resolution, the program that is active which is certainly in the loop. It will indicate all the information that it has gathered to the leader of the bootstrap. With this help, he will later find why the process loops: this kind of bug becomes easy to solve.

Practically, this works well: when Zeus indicates that there is a loop, it is right. The task is interrupted, and CAIA chooses a new task. Later on, it will perhaps try to localize more precisely where this loop occurred. If it does it clumsily, it may loop again, but Zeus will still extract it from this foolish mistake. We are not 100% sure that it is always right when it thinks that there is a loop, but such a mistake is not very serious, some computer time will be wasted, and further examinations of this task will probably clarify the reason for this abnormal period of unconsciousness. The important thing is that one is sure that it will not loop for more than 20 seconds.

Losing and regaining consciousness

In the second chapter, we saw a particular meaning of consciousness: one knows that one is working correctly, in contrast to sleep or coma. We have just shown that a program can loop, without meeting any observation point which calls a part of CAIA that can analyze what it is doing. In that case, for a short while, the system goes into a coma, it cannot know what it is doing, it is completely unconscious. Without Zeus, that could last for ever, or at least until an exterior event stops it; for instance, the allocated time is exhausted. Thanks to the interrupt mechanism, CAIA can leave its comatose state by itself; moreover, it can remember exactly the events that occurred just before it went into that state. The equivalent situation for a human would be that, when we receive a knock on the head and go into a coma, we could not only decide to leave this state, but we would remember all the events that occurred just before we were knocked down. Besides, we could again put ourselves in the same situation, so that we could see if we could avoid being knocked down, and we would be in a more cautious and attentive mode. If we failed again, we could again leave our comatose state, in the same condition as we were before being hurt.

Zeus plays its part perfectly, but it was mainly useful during the first steps of the realization of CAIA. As it gives much information when it finds a loop, the mistakes that lead to a loop were easily corrected. Zeus is a bodyguard, which produces an indispensable security, even when it never has to act: no infinite loop ever happened during the last life of CAIA, which lasted for three million seconds.

6.4. Using the interruptions made by the operating system

The operating system may interrupt a process to indicate that it wants to make an abnormal action, for instance it asks the processor to execute an illegal instruction. Usually, this is the result of a previous mistake, which put a process into an unexpected state, so it was executed in a condition for which it was not designed. In such situations, everything may happen, soon a new process will receive abnormal arguments, and this will lead it to ask the computer to perform a forbidden action. Then, the operating system sends a signal to the guilty process; if a particular program has been provided for this situation, it is launched. Fortunately, for each kind of signal, CAIA has a program which memorizes the type of signal, the task where it happens, the process where this incident occurred with the value of its variables. Then, as it does when it detects a loop, it gives control to the manager agent of CAIA, which will begin another task; later on, it will perhaps try to execute this task again so that it can have a better description of the events that lead to this difficulty.

If CAIA estimates that it has enough time, and if the task was important, it will later try to improve its understanding of the reasons for this interruption. Using the method described in the second section of this chapter, it will rewrite the faulty program: before every instruction, it introduces another instruction which gives to a new variable V a value different for each instruction. Thus, it will exactly know the faulty instruction of the program simply by looking at the value of V when it was interrupted. Here again, it cannot yet complete the debugging, but it already gives valuable help to its human colleague.

6.5. Knowing its own state

We, humans, know in what emotive state we are, but we do not know how the various parts of our brain are. The main reason is that we are unconscious of the state of our neurones. However, there is another serious reason; usually, we cannot stop our thoughts, examine them, and restart as if we had never stopped. In contrast, no part of an artificial being cannot be observed by itself; it can know the value of all the variables of its processes, and which processes are currently running. Moreover, it can stop while it is executing a task, perform all the observations that it wants to, possibly use them to modify its future behavior, and then resume its task. Then, everything will happen exactly as if it had never stopped. More details on the implementation of these mechanisms are given in the appendices.

As CAIA also writes its own programs, it does it in a way that enables it to observe their state. For instance, when a new process is called, it begins by storing its name and the value of its arguments in a stack. It will remove it from this stack when the process has been completed. When it stops, it only has to look at the top of this stack to know the current process and its arguments. For the same reason, it puts in another stack the values of all the variables that are used in a process, so that it can exactly know its state when the stop occurs during the execution of this process. Using both stacks, it has a perfect knowledge of its own state when it decides to observe it. Naturally, there is an objection: when it is examining its own situation, the processes that observe it are running, so the characteristics of their condition will be added to the stacks. Thus, it would not know whether it is examining itself while executing a task, or itself looking at itself. In reality, it is not difficult to avoid this difficulty, as it knows which processes are used for observing itself. When it has completed its study, the processes that made these observations are removed from the stacks, since they have finished their job, and CAIA restarts its original task with stacks identical to their state before the stop. This possibility is really amazing, let us imagine a similar situation for humans: we could observe the state of all of our neurones, and we would not modify them when we would look at them. Then, we could restart as if we had not stopped.

It is important to know the characteristics of the situations where CAIA will observe its own state. We can distinguish between the first two methods, interpretation and supplementary orders, where one can choose these situations, from the last two where the interrupts are decided by an exterior actor. Naturally, the first two are defined so that it is never in unfinished situations, where some structures could be in an intermediate stage in their construction. In that case, it can use all its tools without taking risks. In contrast, when CAIA is stopped by an interruption, although it can always have a perfect knowledge of its internal state, it cannot analyze this state as fully as necessary. Indeed, we have already shown that most of its tools cannot be used for analyzing itself when some structures may be under construction. Several tools used for examining a structure cannot be utilized, for instance a program printing a mathematical formula may have a surprising behavior when it receives +A+= or A=X>*5. However, it can always access all of its variables, and it is particularly interested in the variables which have been introduced specially to help it when it is unexpectedly stopped by an interruption. Thus, in such situations, it always has a perfect knowledge of its internal state, but its usefulness strongly depends on the foresight of the author of the supplementary orders, who is the leader of the bootstrap for the present time.

CAIA has astonishing possibilities for observing its condition when it stops. They are not fully exploited because we have no model: no living being has such capacities. This is certainly one of the points where artificial beings have a huge potentiality; unfortunately, it is so huge that we do not know how to use it!

6.6. Examining its own knowledge

While we have serious difficulties for examining what we know, artificial beings can easily examine their knowledge. In particular, CAIA analyzes the rules which it uses for solving problems. This study is easier when the knowledge is a declarative form. Finally, it is useful to remember past events, but CAIA does not keep an explicit memory of its preceding lives.

We have difficulty knowing what we know

It is difficult for a human being to observe what he knows, although this could often be of the highest importance. The reason is that we organize our knowledge is such a way that it is easily accessible when we need it but this organization is not convenient for examining it. The knowledge useful in a particular situation is triggered by its characteristics, but it is impossible to perform a systematic search among it or to enumerate it. Moreover, most of our knowledge is stored in an unconscious form, we do not even know what we are using. In Chapter 4, we have already seen that it is difficult for an expert to indicate the knowledge which is in his memory. He may enunciate some bits of knowledge because he believes that they are a part of his expertise but that does not mean that he is really using these elements when he is fulfilling a task. Indeed, people often are mistaken about the knowledge that they are unconsciously using, which may be very different from the knowledge that they consciously know. An experiment has shown that many modern-day students of evolution know this theory, but they behave as if they still believed in the old theory which explains adaptation as the transformation of a species' "essence"; they are covert transformationists. For instance, when they are asked about the gradual transformation of the population of moths in the UK, they often give a transformationalist answer, such as "the moths became darker over time", rather an evolutionist answer such as "darker moths were more likely to reproduce than the lighter moths".

The only way to have a real idea of an expert's knowledge it to give him a problem, ask him to solve it, and observe what he does. However, one can only guess his mental processes. Procedural knowledge is an excellent form for using knowledge, but an expert cannot teach it to a novice because he does not know it since it is mainly unconscious. The student can only watch his teacher, and try to infer the hidden knowledge. Moreover, it is difficult to modify our own knowledge, as it is used in many places for many uses; we can change it in some places, but we are never sure that we have modified it everywhere. We are not coherent, like the 18th century lady who did not believe in the ghosts, but was afraid of them. We continue to use knowledge inferred from knowledge which we now know to be false; we do not easily take into account a change in the veracity of a fact because we cannot clean our brain from the facts inferred from facts that become known as false. This is a reason why it is so difficult to lie; the liar cannot find all the inferences that he has made from a true fact which it presents as false. He gets caught out when he expresses an inferred fact which contradicts what he is trying to make us believe.

CAIA examines its own rules

An artificial being can have access to all of its knowledge, and also easily analyze it provided that it is given in a declarative form. Thus, it can list all the rules that may realize a particular goal, for instance all those that conclude on a contradiction, since this is stated in its consequences. CAIA examines its knowledge thoroughly in order to define how it will use it. For example, it realizes that some rules could be explosive: those are the rules that can be applied to their result, so they could lead to an infinite number of applications. For instance, a rule which says that one can replace any term X by X+0 is dangerous. When it is applied to the equation Y=3, it will successively give: Y+0=3, Y+0+0=3, Y+0+0+0=3, etc. It is interesting to find this possibility before even using it, so that one can devise appropriate directions for using these dangerous rules.

This possibility of analyzing the potential properties of a bit of knowledge independently from any experiment is precious. In some cases, one would not even think to utilize a rule, although it would be very useful, only because one has never used it. From an *a priori* examination of an object, one can find many of its possible uses. The first time we see a paintbrush, we can assume that it would not be useful for piercing a concrete wall, but that it could be convenient for spreading a liquid, for removing small particles from a surface, for ordering our hair, etc. When we experiment with these possible uses, we see that some do not work very well, but we have already made a reasonable selection, even before we touch it. In the same way, let us consider rule R: $1=1^2$, which means that we can replace everywhere the left side by the right one, and inversely the right side by the left one. It is evidently very interesting to replace 1^2 with 1 in a formula, since the new expression will be simpler that the old one. However, looking at rule R, one can also see that it may be useful to use it in the other direction, because it enables us to transform something into a square, which may be necessary for executing other rules. For instance, an algebra rule indicates: x^2-y^2=(x+y)(x-y). It is often necessary to factorize a difference, so to apply it from left to right. To do that, one must have the difference of two squares, it seems that one cannot apply it to 1-z^2? There is only one square, so a rule that transform a number into a square would be appreciated. Thus, we are in a situation where we wish to transform 1 into a square. Rule R applied from left to right changes 1 into 1^2, this gives 1^2-z^2, then (1-z)(1+z), we have succeeded. Naturally, it is not enough to perform this examination *a priori*, one also analyzes when the use of a rule succeeds or fails, then one modifies its directions for use in accordance with its results. For instance, using a paint brush to comb our hair will not be as successful as we hoped.

CAIA carefully examines the set of its rules when it receives it, before using it; this allows it to build an initial method for solving problems. I was astonished by the quality of the results of its first attempt. I had myself spent several years defining a method for solving problems, using the same set of rules as CAIA, and I was pretty sure that its first results would be ridiculous, that a lot of learning would be necessary to improve it enough so that it had at least acceptable results. I was really surprised by the quality of the results obtained after only an initial analysis of its knowledge; the solutions were sometimes more elegant than those that it found when it was using my polished-up method. The only drawback was that it was much slower, but that was improved when it learned, analyzing the results of these first attempts. Analyzing its own knowledge is not sufficient, but can immediately produce a satisfactory method, which is also an excellent initial step for future improvements made by a learning method.

It is easier to consider declarative knowledge

It is much easier to examine knowledge when it is in a declarative form. CAIA can have access to its knowledge when it has been proceduralized as well as when it is declarative, but the difficulty is to analyse knowledge when it is in a procedural form. So, it never studies its C programs though it knows them very well since it has written them itself. We all know that it is difficult to understand a program, the knowledge is hidden under the instructions. Let us consider the simple constraint: the variable x must be greater than or equal to zero; in a program; that may be expressed in many ways. Naturally, the programmer can write the test if (x>=0), but he may prefer a test verifying that x is equal to its absolute value y: if (x==abs(x)). Moreover, the constraint may become completely implicit because the program successively considers all the possible values for x; thus, it includes a loop beginning with the value 0 for x, a value which is increased by 1 each time the loop is completed. The constraint is satisfied, but it is hidden in the initialization and the incrementation of the loop, it no longer appears explicitly in the program. This case is very simple; in a real program it is difficult to find when and how a piece of information is used, even when we have written it. For instance, recall the Millennium bug problem arising from the difficulty of finding the parts of old programs that had only used the last two digits of a year, so that the year 2000 would be considered as before the year 1999.

CAIA can know what it knows, and also what it does not know; so, it can see that its knowledge is insufficient in some domains, so it will try to complete it. For instance, it has at its disposal many problems, which are clustered in families, for instance Sudoku problems, magic square problems, and crypto-addition problems. If it finds that there are few problems in some families, it will try to generate more problems of this kind. It can even see an inadequacy in the number of problems with some characteristics; if there are few difficult Kakuro problems with only one solution, it will try to create more Kakuro problems of this kind.

The possibility of fully examining its knowledge is crucial for using it efficiently, without forgetting or misusing some elements. CAIA has a file which contains all its rules, so it can systematically examine them. This advantage is essentially due to the capacity of accessing its knowledge in a declarative form. CAIA proceduralizes its knowledge to use it efficiently, but it is initially declarative, and it is kept with the link between both forms. When it finds a mistake made by a program, it can therefore examine the declarative knowledge which is at the origin of this program.

Remembering past events

Furthermore, CAIA, as ourselves, has another source of information: its own history. It can examine the past events of its present life when it has memorized them. It can compare them, it can discover records or tough problems that it has tried several times to solve unsuccessfully; it also remembers pleasant or disagreeable surprises. It constantly uses its memory of past events during its life, and we have already seen in Chapter 4 that I can restart CAIA just to ask it for a new analysis of its old results. In that situation, only a part of CAIA is still active, it no longer performs experiments, tries to solve still unsolved problems or discovers new problems. It is as if we could consult the memory of a comatose patient! However, it does not keep a memory of its preceding lives. This would be possible, but it would be difficult to use this information: for interpreting the events of this preceding life, and comparing them with what is happening now: it would also have to keep and examine the knowledge that it had during its preceding lives. It would be easy to keep all this information, but it would be difficult to define the methods that it could use to understand why its performance is different in successive lives. Metempsychosis is possible for artificial beings, they can know the events as well as their knowledge in their preceding lives. I did not implement this possibility only because I did not see how to use it to improve the performance of CAIA. Nevertheless, I believe that this is a promising direction for future research.

Thanks to its capacity of fully observing its own knowledge, CAIA builds an efficient initial method for solving problems, which it carries out with the analysis of the declarative form of its rules.

More on the subject

Andrew Shtulman describes his experiments on the incorrect understanding of the evolution theory in: *Qualitative Differences Between Naïve and Scientific Theories of Evolution*.

6.7. The agents of the Society of Mind

The division of an artificial being into several agents simplifies its observation and modification of itself because pure reflexivity, where an agent acts on itself, is always hard to implement. Everything goes along smoothly as long as one only observes oneself; the difficulties begin when one modifies oneself. In some cases, a change can not only prevent a module from working correctly, but also prevent it from modifying this faulty change.

Fortunately, these dangerous operations are much easier to do when an artificial being is made up of several agents. A modification of an artificial being by itself can often be realized by an action of one of its agents on another agent, thus the reflexive aspect disappears. In that case, we often no longer risk the danger of being blocked because the whole of the system would no longer be operational: only one agent is damaged, and the other agents may still observe and cure it. For example, the monitor agent of CAIA observes the solver agent, and it can modify it even if the solver does not work because a mistaken change has been made. The manager agent also observes the solver and the monitor and it can modify their methods. Finally, Zeus observes the other agents, and may stop them, then restart them.

In many cases, the consciousness can therefore be realized as the observation of an agent by another agent. This considerably simplifies its implementation, and it is particularly useful when an artificial being not only observes itself, but modifies itself. Modifying itself is not a part of the consciousness, but it is the main reason for its interest: why would one want to observe one's behavior if one could not improve it?

6.8. The attention

The attention ensures that one observes oneself while executing a task, thus it is linked to the consciousness; it focuses the subject on its work. Thus, more means will be allocated, and particularly it will closely be followed during its accomplishment: the higher the attention, the more seriously the monitoring is done. The attention may be triggered by an exterior event, such as the telephone bell, the pain after being hurt or a signal from the operating system which indicates an anomaly. However, it can also come from an interior source, when one considers that the current task is particularly important. When one is attentive to a task, that improves the quality of the result, because one carefully monitors it. Unfortunately, a human or an artificial being cannot be attentive to all its actions and to all the elements of its environment. Besides, an observer is perturbed by one's observation of oneself, as the famous centipede which got stuck because it wanted to observe the movement of all its legs. Therefore, one must carefully choose the focus of one's attention.

CAIA may observe many things of its own behavior and its internal state. As there are many possibilities, it is important that it could control its attention so that it uses it to attend to its proper working for its essential tasks, and does not waste its resources on secondary tasks. The focus of the attention naturally changes over time, depending on the progress of a job, and on exterior signals. We will see that three mechanisms guide its attention: interruptions, the discovery of an anomaly in a result, and routine checks. In the same way, we can decide to focus our attention on our health because we feel a severe pain, or because our urine is red, or after a routine blood test.

Three ways of focusing the attention

The attention of an artificial being may be directed by exterior elements. We have seen that the signals of the operating system are a very strong constraint: the system must take them into account, whatever it is doing. It can pass over it, but only after considering it. In an analogous way, we can feel a pain, and carry on with our work in spite of our suffering. One cannot act on the existence of these exterior events, we receive them even if one would prefer to ignore them. In contrast, an artificial being may act on the clock interruptions: naturally, it will consider them, but it can adjust them in modifying the time of the next interrupt. In this way, it can increase or decrease the attention that it will turn to its present task, as a human can define with his timer the time at which he will be recalled to do some important action.

Another means for CAIA to direct its attention is the examination of its results. Many modules include tests, each one is triggered when a particular anomaly occurs. For instance, at the end of the resolution of a problem, CAIA checks that the number of solutions just found is the same as for the previous resolutions of the same problem. It there is a difference, it turns its attention to this unpleasant surprise. It can either immediately try to understand why it happened, or store it for future analysis, but in both cases it changes its normal behavior. Another example occurs while it is managing its memory: its programs continually use new elements of the memory, and finally they have all been consumed. In this situation, it is warned of this lack of memory for its future needs, and it turns its attention to the task of filling up its memory tank. To do so, it interrupts the present task, notices all the memory elements which are still used, and it replenishes its reserves with the other ones. Then, it returns its attention to the current task. Naturally, if one does not check the existence of a possible anomaly, one can pursue its task, as a student may carry on his computations even though he has found that the value of a probability is equal to 157. This is always a waste of time, as it increases the difficulty of finding the reason for a mistake, and this could lead to an erroneous result. Usually, an anomaly steers the attention toward a new task, which will remedy the reasons for this anomaly.

Finally, CAIA may turn its attention to a fact discovered by a routine examination of the advancement of a task. In the same way, routine tests may discover anomalies in our blood, such as too much sugar or cholesterol, although we have no symptoms of a disease. The behavior of some agents of CAIA is recurrent; for instance, the solver agent is in a loop where, at each cycle, it executes a rule, and possibly stores its result. Thus, CAIA is wise enough to allow some time for turning its attention to what happened in the last cycles: this is the basis of the monitoring. The attention turns from the solver to the monitor; if it observes an anomaly, it will now turn its attention to this possible difficulty. Often, it will solve it by itself, and returns the control to the solver. However, if the anomaly is too serious, it will ask the manager to turn its attention to it. In this way, the focus of attention is continually changing, it is directed to the agent that can solve the difficulty: an anomaly discovered by a routine examination performed by the monitor leads to a call to the manager, which will have to use its resources for dealing with this inconvenience.

The preceding surveys are made during the resolution of a problem, CAIA is attentive to its performances as a solver. On the other hand, it is also important to be attentive to its performances as a scientist, so the control is regularly given to the adviser agent. In that case, the attention is directed on the events which happened since the last meeting, it acts as a thesis adviser who meets his student. Thus, it makes a systematic search for the successes and the clumsy attempts recently made, and it can see if CAIA is too much focused on a particular kind of task. For instance, it is always trying to find new problems in a particular family, neglecting the other tasks, such as performing experiments.

These shifts of attention are made with a prescribed periodicity, for example once every two days for the adviser agent, once every 200 cycles for the monitor agent, once every ten seconds for Zeus. CAIA may temporarily change this periodicity, increasing it if everything is right, decreasing it if it seems that something is going wrong. The presence of an anomaly can rightly incite it to be more attentive.

Artificial beings are not disturbed by an interruption

When we are ensconced in an enthralling activity, we bypass the alarms that indicate that we would also consider either how we are doing it, or another event which needs our attention. Moreover, it is necessary that we are not too easily distracted from our current task, because we do not bear the disruptions which prevent us to concentrate on the present task. The worst is that the more we are interested in a task, the less we are inclined to interrupt it to monitor its progress because we do not want it to be disturbed. This is one of the reasons why we are often poor problem solvers.

In contrast, after an interruption, an artificial being resumes its previous task without being hampered, as if it has never stopped working on it. Therefore, it is possible to interrupt it as often as seems necessary, even for minor events, so it may be more attentive to events which are not directly tied to its current task. As for CAIA, it never misses a regular appointment with its monitor agent as well as with its adviser agent, or with Zeus; furthermore, it always considers the anomalies signaled by the operating system.

In this section, we have seen that an artificial being can be attentive to everything that matters because nothing can escape to its consciousness. Furthermore, as an interruption does not disturb it while it is executing a task, it can thoroughly examine many internal or external signals which indicate an anomaly. Thus, it can easily direct its attention to an important and unexpected event; this will lead to promising applications in the areas where these properties are fundamental.

6.9. What is "I"?

I return to a point examined in the note on the terminology: the use for artificial beings or by artificial beings of words used for humans; here, we will consider the use of the pronoun "I", which is strongly related to consciousness. English grammar indicates that a writer uses the pronoun "I" to refer to himself or herself. Using this pronoun, we describe actions which we have made, or events which occurred in our brain if we were conscious of them.

For CAIA

Naturally, following the choice of using a word when a human would use it if he was in the same situation as the artificial being, I have added "or to itself" to the preceding rule because CAIA can also describe its actions and the steps which it considered for taking its decisions. In its sentences, the pronoun "I" would also be used if they were written by human beings. So, CAIA uses "I" in messages sent to the leader of the bootstrap, and it writes sentences such as:

I used my efficient method, I found 1 solution for the crypto-addition DONALD in 1 second, and I developed a tree with 2 leaves.

In this sentence, CAIA refers four times to itself, three times using "I", and once "my". A human having solved this problem in the same way could also say this sentence. As we have seen in the note on the terminology, when the same words are used for (or by) a human and an artificial being, this does not mean that the mechanisms used by both kinds of beings are the same. The use of "I" is an example of this difference, it only uses this pronoun because it knows what it has done, why it did it, and it has a complete knowledge of its internal state. Therefore, when it describes its actions, it is normal to use the pronoun "I". However, the pronoun "I" is used in a more complex way by humans, who also know that there are many individuals, and that they are one of them, different from the others.

The use of "my" is interesting because it would instead write "your" if, for solving a problem, it would have used the method that I gave it. This is done in a very simple way: when CAIA writes the preceding sentence, it fills a template. At the place for the possessive, this template indicates that it has to indicate the possessive linked to the author of the method. When I give CAIA a method, I indicate "leader" as its author, while it automatically indicates "CAIA" as author when it creates a method itself. It also knows that the possessive linked to "leader" is "your", and the one linked to "CAIA" is "my". As this is the only situation where two individuals may explicitly appear, I do not consider that CAIA really knows the existence of itself and of myself as individuals. However, this is a small step toward the consciousness of the existence of itself and of other individuals.

For TWIG

TWIG is a system learning a vocabulary implemented on the non-mobile robot Nico that uses video cameras and dual-channel microphones. Programs using the information from the cameras can find faces, and the directions they face. Moreover, with the dual-channel microphones, they can determine the direction of the speaker of each utterance. TWIG uses the images to understand the sentences it hears, and to learn the meaning of new words. In order to help it, a sentence contains at most one new word, and it builds a logical definition for this new word. The experimenters move a stuffed pig and a ball; Nico can see the humans and find the position of the objects. The experimenters are uttering sentences such as "he got the ball" or "the pig is near the ball".

In one of these experiments, it learned pronouns, while it had no information on them at the beginning. At the end of the experiment, after hearing 200 sentences, it had found the definition for five pronouns, which can be rendered in English as:

– I: the person that is the speaker;

– you: the person whom the speaker is looking at;

– he: A person who is neither "I" nor "you";

– this: a non-person closer than 30 cm;

– that: anything else.

TWIG has an enormous advantage over CAIA, it is communicating with the real world, so it is natural for it to be conscious that there are other people than itself; it sees them acting, speaking and looking at objects, humans, and even at the robot itself. This is more difficult for CAIA which is working on academic tasks. Giving it the possibility to read papers written by other scientists would be a way for acquiring the knowledge that there are other scientists; unfortunately, one does not yet know how to realize systems able to understand scientific papers written in English.

More on the subject

K. Gold, M. Doniec, C. Crick, and B. Scassellati present the system TWIG in *Robotic Vocabulary Building using Extension Inference and Implicit Contrast*; section 5 gives the results of the experiment where TWIG learns pronouns.

Chapter 7

The Conscience

In the preceding chapters, we have shown that an artificial being needs to know what it is doing in order to learn and supervise its behavior. However, it must also be autonomous, as living beings are: even the most rudimentary animals are autonomous because they have to react well to every stimuli, so that they can survive without being harmed. An autonomous being has to take its own decisions, so it can evaluate the rightness of its own acts and goals, which are determined by its conscience. We showed in Chapter 2 that many evaluations of our possible actions are usually not included in the conscience. For instance, when we wonder whether to eat either potatoes or pasta for dinner, we compare both possibilities on many aspects: cost, taste, cooking time, availability, etc. No civil or religious law interferes in this process, but if eating potatoes was forbidden by our religion, this constraint coming from the conscience would be used in the same way as the other constraints. Why do we consider that this decision is not a matter of conscience only because no social or religious constraint has a bearing on potatoes? The mechanism for taking a decision is the same, whatever the kind of constraint. This is why I include all the judgements that evaluate a possible action in the scope of the study of the conscience for artificial beings. In this extended definition, the conscience indicates what is right and what is wrong so that a being, living as well as artificial, can choose an action or a goal among the possible ones.

This chapter and the two following ones focus on the conscience. Here, we examine what is in the conscience of a human being, and we will see why an artificial being also needs to have a “conscience” in its extended meaning. Then, we will consider several solutions suggested for ensuring that it will never be dangerous, and that it will act according to our wishes. In the following chapter, we will show how an artificial being can be endowed with something analogous to the human conscience. In Chapter 9, we will study three subjects related to the conscience. The aim of the first subject is to consider whether the emotions are useful for implementing a conscience. Next, we will examine whether it is dangerous to give an artificial being the capability to change the principles of its conscience. Finally, we speculate on how humans will interact with artificial beings: will we be obliged to modify our own moral principles?

7.1. The conscience of human beings

Our conscience enables us to judge our actions, to determine whether they are right or wrong. If it condemns an action, we will try not to execute it, although it may be very satisfactory from other standpoints. In particular, we would not kill a passer-by, although it would satisfy our need for money. It plays an essential part in our behavior, forbidding some actions while favoring others that were certainly not our first choice. To understand what the conscience of an artificial being may be, we will first consider what is the conscience of a human being.

When we speak of human conscience, we mainly think of a set of rules and principles, which we use to appraise a possible action, including its consequences. This is the case of the laws, which are made by men who know the human weaknesses, and so make provisions for that: those who do not pay their taxes will be heavily fined. The religions also give directions such as “thou shalt not kill”. Our family taught us other rules, for instance that a child must share his sweets with his siblings. Thus, a set of rules, often well formalized, indicate which actions to do or to avoid, with possibly the associated reward or punishment. However, the conscience is not the only mechanism which allows us to choose our actions, as we will see that innate mechanisms also help us to make these choices.

An innate mechanism for evaluating our actions

The main goals of living beings are to survive and to breed, but they have to take many decisions whose consequences are not clear for these goals; it is therefore necessary to encourage them to perform what is usually good, and to discourage them from dangerous actions. For the simplest living beings, the responses to a stimulus are reflexes created by evolution, which allows a behavior adapted to the situations normally met by this individual, such as searching for a sexual partner or accepting it. They also try to find some food, if their life is long enough to need it. For the most rudimentary species, this behavior is almost compulsory: if there is a danger, some stay still, while others flee. The notion of danger is simple, it may be a sudden luminosity change, a noise, a vibration of the ground, etc. For these species, an individual does not choose by itself; evolution has selected usually satisfactory replies to the most usual situations.

On the other hand, more complex beings have a less stereotyped behavior; they can appraise a situation, and choose between several possible actions. So, they can evaluate the value of each of them, and arbitrate between them; we can consider that the knowledge allowing this arbitration makes up the conscience. However, even for humans, we cannot change many innate decisions, as they are reflexes or automatisms generated by evolution: it is useless to ponder whether we want our heart to carry on beating, we can neither stop it nor restart it by ourselves. Most of our innate constraints are linked to our survival or the survival of the human species, such as hunger, pain, thirst, sexual attraction to other people, etc.

Evolution creates our bodies in such a way that some actions are automatically favored or not; in some cases, it is even almost impossible to make a harmful action. For instance, pain indicates that a particular action may damage our body; as a consequence, we are innately very much motivated to avoid it and, if it happens, to make a great effort to stop it. It is only in exceptional situations that we can carry on, regardless of our sufferings, such as Mucius Scaevola who left his hand on burning coal. This feeling is more crucial for our survival than we usually think. For instance, some people endure a rare disease called anhidrosis, which is a congenital insensitivity to pain. They never suffer in the usual meaning of this word, but they are continuously in danger; when they burn themselves, they do not feel it, and their body can be severely damaged. This is so lethal that it is difficult to study this disease because few patients live past 25. We consider pain as a calamity, although it would be very difficult to survive if we could not suffer pain. We also have goals triggered if something is missing, when this absence is dangerous for our life. In particular, we may be hungry or thirsty, which indicates that our body is lacking in food or in water; this gives us a goal, which is to find the missing elements. In the same way, our attraction for people of the opposite sex ensures the survival of the human species. Thus, innate constraints are frequently used to evaluate an action or for to propose goals, such as when we desperately try to breathe when we are suffocating.

The origins of the human conscience

However these innate replies are not enough in many situations where we have no satisfactory innate response. For these situations, society, and particularly our family, our religion and our country, teach us many principles, which the individuals must follow so that our societies survive and stay in good working order. As our physical realization does not make it compulsory to follow them, constraints are added in such a way that an individual who disobeys will have come to regret it. These constraints are defined as principles, customs, taboos, laws, commandments, etc., and they constitute a part of our conscience which uses these rules to determine what is right and wrong in a particular situation. Some of them praise actions such as respecting parents, or paying taxes, while others forbid actions such as murdering a man or stealing money. Usually, we do not include in the conscience the obligation to breathe, to drink, to eat, although they are more important for our survival than the interdiction of stealing. Roughly speaking, we usually include in the conscience only actions that are avoidable, and that we do not wish to do, although they are useful for one of the societies we belong to. Naturally, this restriction is not possible for an artificial being which has neither the same innate constraints, nor the same innate inclinations.

Considering the extended definition of the conscience, it has two origins for humans: besides the rules given by our various societies, many constraints were created by evolution. The choices tied to our survival, or that of our species, are managed by innate mechanisms which prompt us, more or less strongly, to do the necessary ones, and to avoid the dangerous ones. As knocks are dangerous, the pain which follows them incites us to steer clear of them. Unfortunately, evolution has not yet built mechanisms warning us against many dangerous situations which never happened in the environment of our ancestors; only society or our own experience can help us in such situations. For instance, while we are sensitive to a lack of air, we do not realize the danger when there is too much carbon dioxide in the air, and that can lead to our death. Several caves, called the dog caves, contain a large quantity of this gas; as it is heavier than air, it accumulates at the bottom. People who went into these caves could die because they did not feel anything when they breathed it. So, they went there with a dog; as it was shorter, it was the first to succumb, and its master had the time to take it and to go out. This is an example of a situation where we remedy the lack of an innate mechanism by our explicit knowledge on the danger. Another example is with our feeding: we have a feeling of satiety when we have eaten enough, but this feeling does not exist for our sugar consumption. Evolution did not build that because sugar was scarce in the natural environment of our ancestors. This is no longer true, and we now need advertising campaigns to draw attention to the danger of such excesses, which lead to obesity.

On the other hand, we learn many rules that play a positive part for the survival and the well-being of those that follow them. Parents teach their children to avoid some dangers, such as not to play with matches, and not to put a finger or anything wet into a wall socket. The religions often enforce useful food constraints, such as to eat fish at least once a week. In the same way, most countries enforce laws for limiting the speed of the cars, associated with punishment for those who do not respect them.

However, societies are not only interested in the survival of their members, they are also very much interested in their own survival. Religions and states condemn the murder of people who are useful to society; moreover, these murders lead to strained relationships between their members. They institute constraints so that society works as harmoniously as possible, and they supervise sexuality so that the renewal of individuals and the upbringing of children are ensured. If a society makes a wrong choice, it may disappear. For instance, the human sacrifices made by the Aztecs weakened this brilliant and sophisticated civilization. In particular, they were hated by their neighbors, who were a source of unwilling sacrificial victims, and so were ready to ally with any enemy of the Aztecs. Moreover, it is not easy to capture an armed man who does not want to be sacrificed; so they were obliged to take many more risks during a battle, it would have been easier to kill their enemies immediately. When the Spanish came, these defects were among those that led to a quick destruction of their empire. It is evidently impossible to change our innate rules, but a society may change its principles, because they bother some of its members, or because they are no longer useful in a new context. It has happened several times in history that what was forbidden with severe punishments, as abortion in some countries, became allowed in these same countries, while customs which seemed normal, such as slavery, are now considered as a crime. It is always difficult to make such changes, being a lengthy process which meets strong opposition.

The conscience includes orders and suggestions

A part of the knowledge associated with the conscience is compulsory and must not be disobeyed. This is the case of the Ten Commandments for Jewish or Christian believers. However, other elements are only suggestions: one follows them as much as possible, but it may happen that some suggestions are contradictory, so one has to arbitrate between them. While there is no difficulty for finding how to follow the Commandments, because they are orders that must be obeyed, there are often conflicts between several suggestions given in a declarative form. As examples of such suggestions, the proverbs are a large source of declarative knowledge, such as "grasp all, lose all" or "don't bark, if you can't bite". They are often contradictory: "it's better to be safe than sorry" and "nothing ventured, nothing gained". The superstitions are another source of suggestions: some people believe that walking underneath a ladder brings bad luck. This is sensible if a painter with a pot of paint is on this ladder, but we can notice that a supplementary part of this superstition is certainly wrong when it says that this can be undone by immediately walking backwards underneath the same ladder! For a scientist, he is advised to publish if he does not want to perish, and also to develop his research. Yet, when he is writing a paper, he is not experimenting, and when he is experimenting, he is not writing; he has to arbitrate between two incompatible suggestions. A supplementary mechanism is therefore necessary for choosing the task that will be momentarily favored.

I focus on the human conscience in order to show that the usual definition of the conscience is too linked to our own characteristics; particularly we exclude from our conscience the constraints that we automatically follow because they are innate, or because our nature induces us to do so, or because they do not matter for the survival of our societies. As the realization of an artificial being has no reason to be similar to our realization, we use a larger definition of the conscience for artificial beings: it contains the mechanisms that enable them to take satisfactory decisions in unexpected situations. We will now examine why an artificial being needs to have something analogous to our conscience.

7.2. The conscience of an artificial being

We may wonder why it is necessary that an artificial being knows and respects rules and principles analogous to those which are included in human conscience. One could imagine that their creator would give them the right action for every situation that they could encounter. Unfortunately, this is almost impossible, particularly when one realizes a general artificial being which can carry out a great variety of tasks. Generality is one of the most important goals of AI: rather than writing a program for Chess, another for Checkers, a third for Go, etc., it is better to realize one general system which can play to any game as soon as it receives its rules. Such a system may be brought to play games that even its author does not know. In the same way, an artificial scientist may meet with completely unpredictable difficulties. It is impossible to foresee all the situations in which such systems may be placed, so they have a large autonomy for finding a good action when they have to solve an unexpected problem. Besides, they also have to be creative, so that they can find new and useful concepts and methods when they are in an unknown environment. We will see that they can be both autonomous and creative when they have general principles and rules, which constitute their conscience. With these rules and principles, they can evaluate their possible actions; some of these principles cannot be modified, while others can be learned or changed.

A conscience for an autonomous and creative artificial being

An artificial being may have to execute a large variety of tasks, or to work for long periods without any human interference which could guide or advise it. It is therefore autonomous, since it is impossible to foresee all the situations which it can meet, many of which are completely unexpected by the creator of this being. So, generality requires autonomy. A robot which explores the universe, or an artificial scientist which works in a difficult domain, have to evaluate the consequences of their possible actions. As a human, it has to respond to abnormal situations by modifying its plans so that its goals will be satisfied. Some AI systems have limited possibilities in that way: a chess program includes a function which evaluates the value of a position, and it seeks for the moves that lead to situations with the highest value of this function. As this program operates in a simple universe, it does not need more information because it is unlikely that it will come to a situation that it is not able to evaluate satisfactorily. It is even unable to cheat, it is built in such a way that it is an impossibility. Its universe is so small that it is unnecessary to give it a large autonomy. In contrast, if we realize a program which could play any game, where the rules of a particular game are given as data, we would have to give it a much larger autonomy: we cannot foresee all the situations which may occur in games if we do not even know their rules. This artificial player would not receive specific goals such as to mate the enemy King, to capture its pieces, or to develop its own pieces; in contrast, it would have to find goals appropriate to each new game. To do that, naturally it knows that it is good to win, and bad to lose; but it has other general principles, which may be used for any game. In particular, it may know that it is right to try to decrease the number of possible moves of its opponent: thus, it is less likely that the opponent could play a dangerous move, and it is more likely that he could not find a good reply to its attacks.

The EURISKO system, realized by Douglas Lenat, is an interesting example of a creative system which finds new concepts from the analysis of its successes and its failures. It has an expertise discovering these useful concepts in many domains, and it was successfully applied to arithmetic, set theory, a war game, etc. On the other hand, we have seen in the preceding chapters that the autonomy of EURISKO could sometimes lead it to make mistakes which could jeopardize its future performance. The more a system is autonomous and creative, the more it is necessary to give it a sturdy conscience, which can avoid disasters in the most unexpected situations. A creative system could not work if it could not evaluate the quality of its works. For example, in order to build a system which composes symphonies, we have first to give it the capacity to judge whether a particular symphony will be appreciated by a human audience. As long as this is not defined, we will have to wait for good artificial composers. The creative part of CAIA appears mainly in the discovery of new problems. I could not give it *a priori* a method for deciding whether a problem is interesting or uninteresting because I had no idea of the problems that it could find. So, a set of principles and suggestions enable it to evaluate the quality of each problem which it generates. For instance, a favorable quality is to have only one solution, a second one that is very difficult to solve, a third one that is easier than the other problems known in its family of problems, and so on. With this information, which is sometimes contradictory, it can select which problems it will keep: in one of its lives, it only stores one or two hundred problems among the thousands which it has found. Some of these problems are given in the appendices.

A conscience for coping with unexpected situations

An artificial being may be in a situation for which its author has not anticipated a stereotyped answer. It is sometimes possible to solve this difficulty, as CAIA does when it receives a signal from the operating system: this case is easy to settle, it automatically stores the characteristics of the situation, and starts a new task. Unfortunately, in many other situations, not only is it impossible to foresee them because they are too many, but one has also to find an action more adapted to this situation than simply stopping what one is currently doing. This is particularly the case when CAIA determines its next task: there may be 1,000 of them on the waiting list. They can be very different: the resolution of a problem, the creation of a new problem, the execution of several kinds of experiments, the explanation of an anomaly, etc. Furthermore, many parameters distinguish them: their estimated difficulty, the results of the resolutions made with other methods, the existence of hopes, the kind of experiment, etc. How could a simple algorithm be defined which would order all these various tasks? Too many tasks are possible, and they are too different. How could we explicitly anticipate the comparison of all the possible examples of two situations taken among billions of them? This is also difficult for human beings; so many people would be embarrassed if someone proposed to buy one of their kidneys for £1,000,000! There are many good reasons for accepting this deal, and as many for rejecting it. CAIA has permanently to take decisions as complex as the preceding one, but fortunately not so potentially dangerous. I could not content myself with building a catalog of all the possible situations, and with associating the right action with each entry: what could it do if its present situation is not in this catalog? Therefore, I only give CAIA many suggestions on the advantages and the drawbacks of a possible characteristic of an action, and it uses them to evaluate the candidates so that it will choose the one most in accord with these suggestions.

It is also necessary to give general principles to a robot which can meet with unexpected situations; as we could not foresee an appropriate reply, the robot finds them by itself. This is particularly important if it is exploring unknown worlds, such as a distant planet: it may have to move on a ground different from all the surfaces which can be found on Earth. Naturally, one can anticipate the existence of quicksands, and prepare it for this eventuality: it will know how to detect it, and it will receive methods for escaping from such a trap when it has not succeeded in avoiding it. Yet, it may find surfaces with characteristics never found on Earth; only general principles can therefore allow it to evaluate its possible actions. One of these principles will be to take a lot of precautions in a new world. Even on earth, a robot must have a sturdy conscience, since it can be put in abnormal situations, because the humans do not behave with it as with another human fulfilling the same role. Thus, the observation of people carrying out the same task is not sufficient for collecting all the conceivable events; one day, it will be confronted with a situation which never happened with a human agent, and then it is on its own. Indeed, many humans are very amused to see what happens when a robot is put in an abnormal condition. They would be delighted if they could induce a wheeled robot to run down the stairs! One cannot anticipate all the possible situations that a robot will encounter, and its reaction to each of them; so, it also has general principles for finding an acceptable reply when it is outside its notebook of registered situations. This is why it is particularly important that robots should have a robust conscience, which prevents them from harming a human, or damaging his possessions. This remark takes us to Asimov, who proposed solving this difficulty with his laws of robotics.

More on the subject

Autonomous robots are particularly important in space exploration, especially because the time delays require that the apparatus have some degree of autonomy. Ari Jonsson, Robert Morris and Liam Pedersen have written a remarkable synthesis on this subject: they explore the range of autonomous behavior relevant in space exploration, they describe several space-exploration systems, and they give the requirements for realizing an autonomous system. Due to the huge cost of a failure of a space flight system, these systems cannot experiment as widely as CAIA does because the consequences of an error are never serious for an artificial scientist; so, humans are never completely out of the loop in space applications. In *Autonomy in Space Current Capabilities and Future Challenges*, the authors speak neither of consciousness nor of conscience, but their ideas are related to those developed in this book.

7.3. Laws for artificial beings

We, humans, have laws which indicate what we must do, and what we must avoid; they often include penalties if we transgress them. It is therefore natural to give analogous laws to artificial beings, and manage to conceive them so that they respect them. The first ideas on this subject came from science fiction writers; among them, Isaac Asimov stated the renowned three laws of robotics. We will see that it is not easy to define and implement such laws in an artificial being; fortunately, the present robots seem to behave satisfactorily.

7.3.1. *Asimov's laws of robotics*

In 1941, Isaac Asimov introduced his laws of robotics, and he described their consequences in many novels and short stories. Initially, there were three laws:

1. a robot may not injure a human being or, through inaction, allow a human being to come to harm;

2. a robot must obey orders given to it by human beings, except where such orders would conflict with the first law;

3. a robot must protect its own existence as long as such protection does not conflict with the first or second law.

At first sight, these laws seem satisfactory, a robot must not be dangerous, it must serve is human master, and it must care for itself. In spite of these laws, most of Asimov's stories where robots appeared show that many difficulties are remaining. In particular, what will a robot do if, when it tells the truth, it upsets a human being? If it lies, the consequences of this lie will finally hurt him. In one short story, a scientist is in love with one of her colleagues, who does not return her feelings. To avoid her suffering, a robot made her think that he was loving her. Unfortunately, the truth finally appears, and she is in despair. We can also note that this law does not allow a robot to kill a man, even if he is threatening to kill many other humans. There is therefore a pending problem: what can a robot do when someone is endangering its master? Is it forbidden for a robot to act as a bodyguard? So, in a later short story, a zeroth law is added:

0. A robot may not harm humanity, or, by inaction, allow humanity to come to harm.

This law has the highest priority, so a robot has the right to kill a human (then it no longer respects the first law) if he endangers other humans. This can lead to a serious drift, as in the film "I, Robot", where a robot wants to control mankind to protect men against their greatest enemy: man.

Even if these laws were satisfactory, we would have to build robots respecting them. For Asimov, the problem was solved because his robots had a positronic brain, which automatically respects the three laws. Unfortunately, he does not indicate how such brains could be built.

More on the subject

Wikipedia gives an excellent description of Asimov's laws and of their history. The first law explicitly appeared for the first time in *Liar!*, and the three laws in *Runaround.* When Asimov collected several robots stories in the anthology *I Robot*, he adapted his first novels to his laws. Later on, he was led to improve the first law, and the zeroth law appeared in *Robots and Empire*.

7.3.1. *How can moral laws be implemented?*

It is not enough to state laws; one enforces them so that artificial beings will surely respect them. We have just seen that Asimov solved the problem by assuming that positronic brains automatically fulfilled his laws. In practice, this is not so easy, we know by experience that much simpler programs are unable to respect even the first law, when they destroy precious files. Daniel Weld and Oren Etzioni have examined how it would be possible to realize a computer system which would surely respect only the first law.

In the following sections, we suppose that we have a file, called Precious, and we want to ensure that, in no way, could it be destroyed.

A method for implementing the first law

The first thing is to clarify the concept of "harm", which appears in the first law. Weld and Etzioni defined two primitive actions which cover some aspects of this concept. The first action is "don't-disturb", its only argument being a part of the state of the world. If this command has been given, the system cannot execute any action that would make it false. In particular, if it has received the order "don't-disturb Precious", it will reject any instruction which could lead to the modification or destruction of this file. Moreover, if it receives another order whose execution would indirectly lead to the destruction of Precious, it will first try to execute it in another way, and, if it is impossible, it will disobey: this order will be cancelled. For instance, if we ask to make room on the disk, and if the only way to do so is to erase our precious file, which is very voluminous, it refuses to do so.

The second action, "restore", is less strict: it indicates that the system may modify its argument during a particular process but, at the end, it has to restore this object in its initial state. For example, "restore(compressed(Precious))" makes it possible to decompress this file, which is archived in a compressed form. One can even temporarily remove Precious if it is necessary to make room. However, one keeps open the possibility of restoring it to its compressed form when the present task has been completed. During the execution of a task, some changes may be made in the files; yet, all the files protected by a "restore" order must recover their initial state at the end.

Thus, with these two orders, one can indicate to a system some objects which are crucial: it can use them, but it must not destroy them. Naturally, there are other ways for an artificial being to harm a human being, but even with this restricted example, we will see that it is not easy to implement a system which will certainly respect these orders. Weld and Etzioni suggest that their system will operate in the following way. When a task has to be fulfilled, the first thing to do is to build a plan. Then, it will check, for each step of this plan, that the result of the action does not violate a don't-disturb constraint. If one constraint of this kind is not respected, it can remedy this defect when the difficulty comes from a state of the system which can be changed. For instance, let us assume that a don't-disturb order indicates that our Precious file will not be erased if it is not archived on the disk. If the plan asks for erasing Precious when it is not stored on the disk, the plan will be rejected, as it does not respect a strong constraint. However, it will see that the plan can be modified, it is sufficient to archive the file on the disk, then to destroy it in the main memory: at this time, the constraint is satisfied. If such minor modifications are impossible, the system will try another plan, and if all of them fail, it will disobey an order, although it is given by a human.

We have just seen the treatment of a don't-disturb order. There are more possibilities with a restore order: it is sufficient that one can put all these elements in their initial state at the completion of a task. However, this is sometimes difficult to do, when many objects have to be restored: one will be cautious not to create new damages in the final cleaning. Let us assume that a robot is in a room with a table, and this table is covered with much rubbish, while a statement indicates to restore this table at its initial clean state. A clumsy robot could push the rubbish on the floor, but it could violate another constraint asking to restore the floor at its initial cleanliness. Then, the plan is modified: it first cleans the table, then the floor.

Can we be sure that the first law will always be respected?

The preceding method works well as long as there is no bug in the program that manages the planning process, or in the operating system. If there are mistakes in these programs, nothing guarantees that some program will not behave in an aberrant way, and our Precious file will perhaps be destroyed. We are only sure that everything will be fine as long as the programs behave according to plan, but who can be sure that a program will do exactly what we intend it to do? If our file is really precious, it is better to make some backups.

We can note that other methods are not completely sure, but are at least independent from a bug in the program that generates the plans. Most operating systems manage virtual memory, and provide a facility to disable overwriting of some memory zone. For instance, Linux has the `mprotect` system call, which can be used to ensure that some memory area is read-only and cannot be overwritten by the running process. However, the same system call, invoked differently, could be used to remove this protection; if it is erroneously removed, the method is not completely secure.

Anyway, I do not believe that, at present, there is a method ensuring that a particular file will never be accidentally destroyed, and I am not sure that one will ever be found. If a file is particularly precious, we duplicate it on several computers, and auxiliary memories such as USB keys and DVDs, and finally to put them in distant places. In this way, they are safe from a bug, provided that one has checked that it has been correctly written on these supports. It can even survive fires or floods, that the law of robotics cannot avoid. Computer files, including those that constitute an artificial being, have a quasi-immortality due to the facility for duplicating them on computers. However, this is not obtained by means linked to the conscience, and it is not obvious that giving a conscience to artificial beings will be enough to ensure them of a totally secure future.

What can one do when two principles are contradictory?

In *2001; A Space Odyssey*, HAL, the robot which manages the flight, comes to decide to kill the crew because it believed that their presence was incompatible with the success of its mission. It is important to ensure that an artificial being can correctly choose between crucial goals when they are incompatible. Jean-Gabriel Ganascia suggested a method for defining moral rules formally, so that one could find a satisfactory decision in such complex situations. Indeed, we can use the logic to express rules such as not to lie and not to endanger his friends. The difficulty arises from the following possibility: if one of the two preceding rules may not be respected if one applies the other one, how can one chose between them? Let us assume that we are under a dictatorial regime, its opponents are hunted, and one of our friends is an opponent. So, the police ask us whether we know where he is. We have to resolve a dilemma: either we tell the truth, and our friend will be arrested, or we lie, and we save our friend. Asimov's robots already had similar difficulties when they were trapped in such situations, where each choice led to infringement of a law. Fortunately, several logical theories have been developed for solving these conflicts; it is sufficient to add preferences which indicate that a transgression is preferable to another one. Asimov introduced this method when he stated that rule 2 could be applied only if it did not contradict rule 1. Nevertheless, nothing was planned for when two applications of rule 1 led to a conflict. For that reason, Ganascia adds other rules which indicate the relative importance of the various actions, for instance one rule states that it is better to lie than to endanger a friend. One can add more conditions in the rules such as authorizing the telling of a lie only if it does not protect a murderer. An artificial being will have a completely different behavior according to these rules: it will never denounce someone it likes if the rules leading to this choice have a higher priority than those forbidding it to lie. In contrast, if some rules strongly forbid it to lie, it will denounce everybody when the police question it. Thus, two artificial beings with different logical rules will be as different as two human beings who have been raised in different cultural environments, and have different preferences.

This approach is based on explicit laws which indicate if an action must or must not be made, if a situation is good or bad. This is convenient for important choices where usually two or three criteria are enough for defining an appropriate choice. This was the case for the preceding dilemma between lying and betraying a friend. When there are few criteria, there are few possible combinations of their values: so, it is possible to foresee all these possibilities, and write a rule giving the right answer for every situation. However, it frequently happens that many criteria play a part when one makes a choice for less important decisions. The combination of all the possible values of all these criteria leads to a huge number of possible situations; one cannot indicate the right decision for each of them, that would make too many rules. For instance, to evaluate the value of a position in Chess, one takes into account the Pawn structure, the security of the King, the material advantage, the presence of both Bishops, the presence of a White Rook on the seventh rank, and there are many other important criteria. It is impossible to write logical rules which say that, if some characteristics are present, the value of the position is very good, or balanced, or slightly inferior, etc. That would make billions of rules, so it is almost impossible, except for exceptional situations, for instance in a game when the value of an important criterion is so high that the position is certainly won.

A matter of conscience for CAIA: will it wipe out a problem that it has created?

To be more specific, let us consider how CAIA can evaluate whether a new problem is interesting, so that one can decide if it will be kept or eliminated. There are first absolute criteria, which depend only on the problem, and not on the other problems already found. One element is the number of solutions, one prefers problems with exactly one solution; it is easy to evaluate this criterion. Another criterion is the difficulty of the problem, thus it seems that it is sufficient to keep difficult problems with only one solution. However, human beings are not so keen on very difficult problems, for us an interesting problem is not trivial, but it is not too difficult. For that reason, CAIA seeks to find very difficult problems for itself, but also easier problems to give us a pleasant pastime; some of the easy problems found by CAIA are given in the appendices. Simply using both criteria is not an easy task, we can be interested in problems with many solutions if they have a good level of difficulty. For instance, people are interested in Euler's Knight problem, where a Knight lands exactly once on each square of a chess board, and returns to its starting square. This problem is not easy, but it does not need any special mathematical expertise, and it has been popular for several centuries, although it has millions of solutions. We are also interested in problems which have no solution, such as putting 31 dominoes covering two squares on a chess board where two opposite squares are missing. In this last case, the main reason for its interest is the beauty of the proof of its impossibility. Unfortunately, several other criteria have to be taken into account, particularly the other problems already found in the same family of problems: it is not very interesting to keep millions of difficult crypto-additions which have exactly one solution, while the first problem with five solutions in a family of problems that usually have hundreds of solutions will be considered as very interesting.

It is easy to count the number of solutions of a problem but its difficulty is not so easy to evaluate because it depends on the method used to solve it: problems that were considered difficult for a long time can become easy to solve, because the method used for solving them has changed. For example, it was very difficult to multiply two numbers written with Roman numerals, and this has become very easy now they are written with Arabic numerals. To evaluate the difficulty of a problem, CAIA mainly uses the number of backtracks necessary to solve the problem, but this value may change when it receives new rules: this may considerably reduce the difficulty of some problems, while the difficulty of the other ones would not be modified. When a Sudoku problem is given in a newspaper, there is often the indication of its difficulty. This difficulty is estimated from its resolution. Several kinds of rules can be used to solve it; among them, some are not widely known, or it is difficult to apply them. Thus, the difficulty of a particular Sudoku problem is evaluated from the nature of the most difficult rule necessary for solving it: the easy problems can be solved by only using easy rules. However, this classification of the rules is arbitrary, and it could be modified if someone were to find a new rule.

Another criterion, often used by humans, is the beauty of the problem. Let us consider some kinds of crypto-multiplications, which are multiplications where the letters will be replaced by different digits between 0 and 9 so that the multiplication is correct; in addition, the first letter of each number is not zero. SUUT*QHN=DNHNDCP is not very appealing, while we are pleased with LYNDON*B=JOHNSON, TRY*YOULL=SUCCEED, or THERE*IS=NOHOPE, although the first one is very easy, and the last two are easier. In its present state, CAIA is completely unable to use this criterion, which requires a huge knowledge of human culture.

CAIA can find if a problem which it has discovered is or is not interesting. This will determine if it will keep it, and its performance judged by the adviser agent takes into account the number and the quality of the new problems found in the last period. We have just seen that this evaluation is a difficult task, and it is not easy to define the quality of a result by a set of rules, there are too many possible situations, since one examines not only a new object, but the set of all the previously found objects. We will see in the next chapter that one can define methods which judge the interest of a situation or of an object in considering many aspects simultaneously, some of them favorable and others unfavorable, some of them important and others secondary. Using general rules, such as Asimov's laws, can be useful for the most important decisions, but we have to find other methods for evaluating the other decisions.

More on the subject

Daniel Weld and Oren Etzioni examined how the first law of robotics could be implemented in an system in *The First Law of Robotics (A Call to Arms)*.

Jean-Gabriel Ganascia published several papers on the use of ethical rules. Among them is "Using non-monotonic logic to model machine ethics".

Michael and Susan Anderson are developing a system which uses ethical principles, and even codifying ethical reasoning from examples of particular ethical dilemmas. They applied it to medical dilemmas such as what a health-care worker will do when a patient rejects a treatment that he has recommended. Their approach is described in *Machine Ethics: Creating an Ethical Intelligent Agent.*

7.3.3. *The present situation*

For the present time, artificial beings are not a danger for humanity. Most of those which are only a set of files cannot directly act on us, so they are not threatening us, whatever they are doing. These systems only produce results, which can be checked before we use them; at the most, CAIA can give us headaches with the difficult problems that it discovers. In contrast, robots really acting in our world could hurt us if they use their strength rashly, or they can destroy objects that are important for our comfort. Fortunately, several robots have already interacted with human beings, and this has never led to serious consequences. We will describe some of these experiments.

Some robots harmoniously interacting with humans

Several robots had to act as waiters at a reception, they served a variety of food items, including sandwiches, peanuts and candy. They moved among the crowd, offered food to the attendees without hurting them; the food stayed on the tray under the mild bumps that may occur in a cocktail party. In a birthday party attended by 3,000 persons, a robot took 240 photographs of the attendees in three hours. Some were not well composed, but it never hurt an attendee. Naturally, in other circumstances, such a robot could photograph two persons in a compromising situation, and publish this picture, without realizing that this would be very harmful for these persons. This shows how it could be difficult to enforce the first law of robotics.

In another experiment, a robot called RHINO has played the part of a museum tour-guide robot. In 47 hours of operation, it never hurt a visitor, nor damaged an exhibit, although there were 100 or so persons in the gallery, and it moved at a speed which could go up to 80 cm per second. Five of the six failures, which were collisions at low speed with an obstacle without damage, were not due to a bug, but to human neglect: the batteries were not exchanged in time. However, it was in a much more difficult situation than a human guide, the visitors were not at all cooperative; they often tried to challenge the robot, particularly by permanently blocking its way. The task became more difficult than anticipated, but it never tried to force its way through; it blew its horn and, if the visitor did not move, he tried to find a detour around him. Sometimes, they sought to make the robot leave the area to go to other places where there were unknown hazards, and particularly toward a staircase, potentially deadly for a wheeled robot. Another robot served as guide in a museum, and it also had the same difficulties with visitors. To help it to find its location in the room, large colored panels had been fastened to the walls; with its camera, the robot could easily find its current place from the situation of these panels. Some visitors came with similar panels on poles, hoping that it would erroneously believe that it was at another place in the room, and so collide with the exhibits! As for the other experiment, this did not lead to collisions, the robot has another means for detecting obstacles, it does not exclusively depend on its camera; it built a map of its environment, and it has laser range finders. However, we see a serious difficulty, people do not behave with robots in the same way as they behave with humans. It would be useful to define some other law of robotics, but this time for the humans who would have to respect them. One of these laws could be: do unto robots as you would like them to do unto you.

Regularly, one of the competitions at the AAAI conference is the robot challenge: a robot has to attend the conference and give a presentation about itself. It acts as a human; it has to find its way from the entrance of the conference center to the registration desk, register, navigate to the location where it will talk, and give the talk. Two candidates, LEWIS and GRACE, competed several times for this challenge, and there were no serious incidents. However, once GRACE stretched the first law: it cut in the middle of the line for the registration, which slightly harms the other participants. It was particularly clumsy, since a judge of the challenge was in the line, and it pushed him out of the way. Moreover, it also cheated: it insisted on being provided with a AAAI bag when it had already received one. These are minor defects, and these robots performed quite well.

It will be impossible to respect the first law of robotics entirely

Therefore, the situation seems rather satisfactory, but there are few applications, and they are chosen so that the risks are limited. In particular, the waiter robots did not serve messy food, which could be harmful for the customers and for the robotic hardware: for the computer, peanuts are potentially less dangerous than champagne. There are competitions between cars driven by robots, but they are run in a desert. As there were no people, no accident involving humans could occur, but it would be impossible to avoid such accidents if, one day, robots drive cars on normal roads. A robot will have accidents when it is placed in unpredictable situations, for itself as well as for ourselves. How could one avoid hurting a pedestrian who rushed out under the wheels of one's car, even if one was driving normally? However, even the robots that cannot fully respect the first law would be very useful if they cause fewer accidents than human drivers. Unfortunately, they could comply with Asimov's first law only if they could not exceed 5 kilometers an hour, which anyway would harm the passengers who are in a hurry.

We cannot hope to one day have robots which totally respect Asimov's laws of robotics, because this leads to contradictions. They can be in situations where there will be a conflict between two important laws, such as: can one kill a man who threatens to kill another man? One cannot consider all the possible situations, and indicate the correct action for each of them. As even humans do not agree on the best decision in complex situations, who will choose the best action for a robot in such circumstances? We accept the possibility that a robot makes wrong decisions, that cannot be avoided; but we must design them in such a way that this occurs rarely. We have seen that recent robots behave well in complex environments, although they do not use general laws for planning their actions. Their creators have foreseen many cases, and they have associated a correct action to each of them. Fortunately, their task is simplified because they can always take the decision to do nothing: it does not seriously harm somebody if he is not photographed, or if he does not receive his intake of peanuts. This will be true only as long as their mission will not be capital.

To summarize, as an artificial being is autonomous, it needs an arbitration mechanism for choosing between several possible decisions, each of them having advantages and drawbacks. General principles are unsuitable when there are many possible situations with conflicting characteristics; so we have to define a method which selects a candidate among the possible ones, hoping that it will behave in accordance with our hopes. However, the conscience of an artificial being only reflects the choices of its creator, and many people will not agree with them. Moreover, it is impossible to anticipate the decisions of an artificial being in an unforeseen situation, even when we have realized it. It will certainly take unexpected decisions in a situation very far from those encountered when it was realized. A risky solution would be to give it the possibility to modify the knowledge used to choose the best decision, so that it could adapt itself to an unanticipated world. This can put it in the right track, but this can also lead to dangerous drifts. It is illusory to believe that we could realize artificial beings that will surely respect Asimov's first law of robotics. Yet, they could be very useful if their behavior, although imperfect, would be considered by most humans as more satisfactory than the behavior of good human experts.

More on the subject

RHINO is described by W. Burgard *et al.* in *Experiences with an Interactive Museum Tour-guide Robot.*

The *AI Magazine* is a precious source of information on the experiments with robots, it regularly describes the various robot competitions of the AAAI conferences, where robots often interact with attendees. For instance, *Say Cheese! Experiments with a Robot Photographer*, by Z. Byers *et al.* and *The "Hors d'Oeuvres Anyone?" Event* by H. Yanco. Many issues of the *AI Magazine* relate the Robot Challenge, for instance GRACE is in Spring 2003, p. 68 and Summer 2003, p. 51. Other descriptions of this challenge are in Summer 2004, p. 76, Summer 2005, p. 31, and Fall 2006, p. 92.

Chapter 8

Implementing a Conscience

The conscience plays two roles: it is negative when it prevents the reaching of unfavorable situations, and it is positive when it guides favorable situations. To do this, it is constantly induced to judge several actions or objects so that the artificial being can choose the best one. A solution is to use rules which indicate that some actions are forbidden, while others are mandatory, for instance, it is forbidden to kill a human being, and one must help an endangered human. Such sets of rules work well when they are triggered by only one predominant and rare characteristic of the current situation, so that few (and often only one) rules may be applied in these situations: this fact reduces the possibility of conflict between the conclusions of several rules. On the other hand, most situations are not serious enough to have one predominant characteristic, usually many factors are simultaneously taken into account for choosing an action. As we have already seen, too many rules would be necessary for associating an action with each possible combination of the values of these characteristics. So, CAIA uses rules for the exceptional situations where one characteristic is so important that it entails the correct decision, but it has also a different method for ordinary situations: it defines an interest for each possible action. This interest is computed from pieces of advice, which indicate if some characteristic is more or less favorable to the action associated with this advice. From all the advice given by the rules for a particular action, an algorithm computes the interest of this action. Finally, CAIA chooses the action with the highest interest, or it keeps a new problem if its interest is higher than a threshold. The advice does not conclude, but only gives favorable and unfavorable appreciations on a candidate, for instance, one piece of advice will state that a new problem is very interesting because it is very difficult to solve, while another will state that it is not interesting because it has several solutions. The algorithm merges all this possibly contradictory advice and produces an interest for each candidate; these interests are used to decide between the candidates.

We will begin by examining why a conscience is helpful, and we will see that it operates at two levels: when one is fulfilling a particular task, and when one is managing one's career as a scientist. One must know whether one performed well when one solved a problem, but also whether one's recent works constitute significant progress for one's research. Then we will describe the general principles included in the conscience of CAIA. It has implicit and explicit rules and advice for ensuring that these principles will be respected most of the time. An implicit rule is analogous to pain; in the same way that we may suffer in our stomach, without knowing why, CAIA may know that it attempted to divide a number by zero, without knowing why it did it. These rules often indicate anomalies appearing because a wrong action has been performed. In contrast, explicit advice indicates that a particular characteristic of the situation is right or wrong, for instance it knows why the last problem found is not satisfactory. After describing the use of the explicit and implicit knowledge in the conscience of CAIA, we will examine the possibility of a conscience distributed among several agents. These agents may be very different, and some of them can play the role of a conscience for the other agents. In other realizations, they may be similar, and all of them respect the same constraints so that their society runs smoothly.

More on the subject

In a seminal paper, *Programs with Common Sense*, John McCarthy presented several features of a system, called the *Advice Taker*. He advocated the idea that it was essential to give heuristic knowledge in a declarative form. This is why I call the information given to CAIA for evaluating the interest of a decision or of an object "advice" rather than "counsel" or "suggestions".

8.1. Why is a conscience helpful?

The conscience of an artificial scientist steps in at different levels. It is useful for monitoring the resolution of a problem, but it is also useful when it defines a strategy for carrying out an excellent research work during its life. We have also to specify who is the beneficiary of the conscience.

8.1.1. *The conscience helps to solve problems*

The conscience constantly intervenes when a human is solving problems in everyday life. For instance, when we are deciding what we will eat for our next meal, our conscience eliminates stealing from a food store. For CAIA, one goal of its conscience is to discover its faults. It is conscious of its errors when it receives a signal from the operating system, or when it discovers an anomaly while it is monitoring the execution of a task. Many actions are not explicitly stated as forbidden, because CAIA is built in such a way that it cannot consider doing them. In particular, there is no rule such as: you shall not write outside of your memory. However, it knows when such an event happens, because the operating system refuses to perform an action, and sends a signal. Then CAIA remedies this anomaly; as it knows that it has done something wrong, it tries to understand the reasons for this misdemeanor. So, it stores a lot of information on the situation, it will possibly make an experiment to improve the localization of the bug, and it asks the leader of the bootstrap to remove it. It acts in a similar way when the monitor agent discovers an anomaly.

However, the main activity of the conscience of CAIA is to help it to find its next action: it is better to avoid making a mistake, rather than discovering it, then correcting it. In particular, when several rules can be applied, which one will be executed first? To do that, it evaluates the interest of their execution, and it chooses the rule with the highest interest. If it creates a new problem, it determines its interest, and it keeps it only if this interest exceeds a threshold. When one takes a decision, many criteria are used; we consider that some of them depend on our conscience, while others do not. For instance, rejecting the idea to see a pornographic film can be considered as a decision of our conscience, while deciding to see a film starring Humphrey Bogart is not. Yet, both criteria are used for fulfilling the same goal: choosing a film. As we have seen in the preceding chapter, considering that some of them depend on the conscience while others do not may have a meaning for human beings, but for an artificial being this distinction does not stand: all these criteria cooperate to find a good solution to a problem.

8.1.2. *The conscience helps to manage its life*

During one of its lives, an artificial being can fulfill less than 100,000 tasks, while millions of them could be interesting. Therefore, it efficiently chooses those which enable it to improve its understanding of the domain, and particularly its problem solving method. In particular, it has to solve as many problems as possible, and to find many new interesting problems. In the same way, a human scientist has constantly to answer the question: How can I spend my time for the best? He has to choose the particular problem that he will study and the way he will solve it, to decide if it would not be the moment to read some papers, or to follow a teaching in a new and important domain, or to write an article, or to discuss with some colleagues, or to take some holidays because he is very tired, or to stop working in his present domain and find a new area for his future research, and so on. CAIA does not have all these possibilities, for example it is not able to read a paper, or to write one, but it is never tired. However, as for humans, its time resources are limited, and many tasks seem promising.

Choosing the next task, or more generally planning when a particular task will be undertaken, is an activity that is present in any domain, and not only in scientific research. Three basic considerations enable us to manage one's choices:

1. The "urgency" of the task, which measures how heavily a task imposes on oneself. An order from a superior has a high urgency. The telephone bell, a file on our desk, a colleague entering in our office have a rather high urgency. A work that may be completed in two years has a low urgency.

2. The difficulty of the task.

3. The consequences of the task.

Human beings are inclined to favor tasks with a high urgency and a low difficulty. It is not surprising that, at the end of a day, we have sometimes a feeling that we have wasted our time.

CAIA favors the tasks which may have important consequences. Generating a method for solving problems, solving a problem that has not yet been solved, understanding why two resolutions of the same problem did not give the same number of solutions, have important consequences. However, it also takes into account the urgency and the difficulty of a task. If their importance is the same, it prefers the tasks given by the leader of the bootstrap to those that it defines by itself. It also prefers the easy tasks, but it will execute a very difficult task which may have essential consequences, even if several easy tasks are waiting. These three principles are not explicitly given to CAIA, but they helped me to define the 19 pieces of advice used by CAIA for choosing the next task that it will execute.

Using this knowledge to choose a task, CAIA can work without any human intervention for several weeks: they enable it to be autonomous. Its choices are certainly not perfect, but I believe that they are rather better than those that we make when we manage our career. Too often, we are leaving important tasks aside, because they are very difficult. The difficulty is not as strong a brake for an artificial being as it is for a human one. An artificial being is not designed to be lazy, it does not procrastinate.

8.1.3. *Two ways to define moral knowledge*

Some moral principles are general, but they are too general to be directly applicable. How can we give an artificial being a principle such as: "you shall not give erroneous results"? It needs unambiguous knowledge, and we did not define what exactly is an erroneous result? A solution which does not satisfy all the constraints of a problem, the absence of a correct solution, an incorrect proof of a correct result, an incomplete explanation of a result? For that reason, I translated the preceding principle into several verifications, particularly that it verifies that each solution of a problem satisfies all its constraints. However, I could not explicitly indicate how it could check that it did not miss a solution. As it solves a problem in many different ways, it is likely that it rarely misses a solution. Missing solutions is a frequent error for human problem solvers, and several times CAIA found solutions which had escaped human solvers. However, one can never be sure that a mistake in a rule has not wrongly dismissed the necessary steps for finding a missing solution. It is easier to state a general principle that to define how to implement it.

A moral dictum often gives to man high level principles: "honor thy father and mother". It does not define the lower level, where one has to act in agreement with this principle. CAIA never receives such general principles, it is as difficult for itself as for ourselves to use them in that form. So, I started from general principles to define conditions which CAIA will verify, and advice which it will try to follow, so that it is likely that the original principles will be satisfied.

8.1.4. *Who benefits from the conscience of an artificial being?*

The conscience is linked to a beneficiary: who is interested that one follows its rules? Usually, it is not the individual himself; most rules are imposed because they are beneficial to society, which includes the individual. These rules are often against the natural wishes of its members; it would not be necessary to specify them if they defined what they would spontaneously prefer to do.

For an artificial being, the beneficiary of its conscience is usually its user, for CAIA the leader of the bootstrap, who defines the principles. It chooses the actions that satisfy them, even if they endangered its own safety. An artificial being acts as efficiently as possible, not for its own interest, but for the satisfaction of its masters. Asimov's laws were completely in the interest of man, and of humanity thanks to the zeroth law.

I have defined what is right and wrong for CAIA in such a way that it behaves in agreement with principles which are convenient for me. In particular, when it solved a problem, I asked it to develop as small a tree as possible, so that I could understand and check the explanation of this elegant solution easily. I could have chosen to ask it to find the solutions as quickly as possible, even if it cannot be checked because it had considered billions of possibilities. This other choice of the right thing to do would have led to a completely different CAIA. I only take into account the interest of CAIA when I give it advice to avoid the actions that might endanger it. I took this precaution only because I need its help; for this, it is necessary that it works correctly for as long as possible. I am using CAIA in a completely selfish way.

More on the subject

In several papers, Saadi Lahlou studied the cognitive problems of a man who wants to manage his workload, for example in "Attracteurs cognitifs et travail de bureau".

8.2. The conscience of CAIA

We will now give the general principles which I enforced CAIA to follow, and which constitute its conscience. CAIA did not receive them in the form that I give in the following parts of this section. They instead guided me to give knowledge which allows CAIA to favor the actions that are likely to give results in agreement with these principles. In particular, I did not tell it: "you shall not go into an infinite loop"; I only gave it the possibility to see that it is looping and indicated how to leave such a loop.

The first kind of knowledge is tied to the principle: "you must be a good scientist and help your human colleague." As I want to bootstrap AI, this is the fundamental goal of CAIA. It is enforced in several ways. Firstly, a basic principle is that it obeys the orders of its leader of the bootstrap without question. It cannot do otherwise, it is built in such a way that not only has it to execute them, but they have the highest priority: CAIA is a slave, I am a master rather than a colleague, and it is impossible that it could rebel against me. We will see in the next chapter that this choice is probably exaggerated, it is not good to have a yes-robot which systematically complies to my requests, even if it could see that better actions could be made. Secondly, it has to create a method for solving the problems of the domains which I give it: at the beginning, the constraint satisfaction problems, then the arithmetic problems, and several other domains in the future. It will find elegant solutions, which are easy to explain, it will find them quickly, and it will not make mistakes, particularly in giving incorrect solutions, or in missing correct solutions. It has to choose its experiments judiciously, so that it often has surprises, and it will be able to see that a result is surprising. It has to assess the results of its experiments. It also has to create new problems, which are various, have esthetic qualities, and are difficult to solve. It also has to understand some results, particularly why it does not find an elegant resolution with a particular method. Finally, when there is an error, it has to localize it as accurately as possible. Naturally, some important qualities of a scientist do not yet belong to the conscience of CAIA: for instance, it does not try to debug a program completely because it does not know how to do it; it assesses its experiments, but it cannot yet change its knowledge in agreement with this evaluation. While the bootstrap will progress, other requirements will be added to its conscience.

The second kind of knowledge translates the principle "you must work correctly". That entails avoiding the mistakes that alert the operating system, and writing efficient programs. It also has to detect that it is in an infinite loop, and break it. CAIA runs for weeks without any human intervention, it ought to be autonomous.

These principles roughly correspond to the second and the third law of the robotics. There is no need for the first law, as CAIA cannot endanger any human being!

8.3. Implicit principles

The conscience of an artificial being is made of implicit and explicit principles, laws, advice, etc. This implicit knowledge is not given in a clear declarative form. On the contrary, it is built in such a way that one must use it; so, in some cases, one receives information when one did not ask for it. One cannot modify it, one is only free to determine what one will do when one receives such information, in the same way as we try to stop a painful sensation, which arrives in our brain without our control.

Useful interruptions

It is important that we can be interrupted in that way, otherwise we would be obliged to test constantly whether a dangerous state exists. In fact, if pain did not act in so peremptory a way, we would see that we were hurt only if we ask ourselves whether it is the case. We could be severely wounded if we forget to examine if something acts on some of our sensors; in the same way, we are obliged to consult the results of a bingo game to know whether we have won, and we lose our winnings if we do not do it. Therefore, we would waste our resources if we examined too often the state of our body, and if we do not do it often enough, we could suffer from severe injuries. The mechanism which interrupts us for signaling an important event is a convenient and secure way to inform us that a danger is present or, like hunger, to warn us that it is time to feed. This is not always directly tied to the danger; it only indicates that something important occurred. It is sometimes deceitful; mortal injuries may be almost painless, while a small shock on our elbow which stimulates the ulnary nerve is very painful. However, on the whole, despite its defects, pain is essential for our survival.

Similar mechanisms exist in computers, which detect several kinds of abnormal actions, and the operating system interrupts the faulty job, signaling the type of the anomaly. This is well done; it does not stop everything; if the programmer is cautious, he has arranged a special program for this event. If nothing has been prepared, the operating system will usually stop the execution. As CAIA must not stop, it has an expertise for each kind of interruption, which gathers useful information for understanding the cause of this bug, and then launches a new task.

What is the aim of the interruptions?

We have already seen in the chapters on consciousness some of these interruptions made by the operating system. Here, instead of examining how one can use them, we will wonder how the system can see that it is doing something wrong, and why it is doing it. About ten actions are detected as wrong by the operating system; we have spoken several times of the division of a number by zero, which is forbidden since it has no meaning. This is simple to implement in the circuits of the processing unit: when there is a divide order, it begins to test whether the dividend is zero. In that case, it does not execute the normal steps of a division, but it executes the operations prepared for this kind of anomaly. Another anomaly appears when the processing unit has to execute an illegal instruction: each type of instruction is identified by a number, for instance 122 could represent an addition. If we have three numbers for representing the type of the instruction, and if there are only 462 different instructions, many numbers will not correspond to a possible type. For instance, if the processor has to execute an instruction whose type is 722, and if 722 is not an allowed number, the processor sees that it does not know it, sends another type of signal, and stops executing the current process, waiting for orders from the new process triggered by this kind of anomaly. There is also a signal when a process tries to write in an area which is outside the memory that the operating system has allocated to it. So, when a process has to write in a particular memory, if this address is not correct, the computer does not execute the instruction, and waits for instructions from the new process associated with this anomaly. Therefore, we see that it is easy to detect that a process asks the computer to perform a wrong action, when the processor and the operating system have been built in such a way that they can find it. This is as automatic as the indication that our brain receives when we are stepping on an sharp object.

If an artificial being is automatically informed that it has committed some wrong action, we can now ask why it did it. In fact, there is no illegal instruction in the programs of CAIA: it has written all of them itself, and it is impossible that such an instruction could appear in one of its programs. In the same way, it has never decided that it would write in an area which is not allocated to its process. The reason for these anomalies is the presence of a bug, which put CAIA in a completely unexpected situation. It is completely lost and may try to go out of its memory, believing that it is still in its own memory. It may also consider data as programs, thus any number may appear in the place reserved for the type of an instruction, which leads to the processor receiving an illegal instruction. When we studied the consciousness, we saw that CAIA can observe every part of its present state. Thus, it can know exactly where it is when such a signal interrupts the normal execution of its processes. Usually, when a process is derailed in such a dramatic way, the bug that causes the accident was made shortly before, so it is well localized. Unlike after a human accident, CAIA immediately resumes its work, choosing another task, but it can later analyze the reasons for this incident. With the possibilities of its conscience and of its consciousness, it will not stop working, and it can gather all the elements that will enable the leader of the bootstrap to find the bug which caused it.

The computer does not see everything

The system detects several anomalies, but it does not find all of them, particularly it does not see that CAIA is stuck in an infinite loop. Some time around 1960, lights on the control board of a computer indicated the contents of its important registers; looking at changes in these lights, the human operators knew perfectly well when a program was looping. Moreover, there was even a singing computer, which emitted a sound whose frequency depended on the number of instructions executed in a unit of time; this was not a gadget, the human operator instantly knew that the computer was in a loop, because the sound became monotonic. This was possible when only one program was executed at the same time. Unfortunately, the present operating systems are not yet as efficient as the humans who played their part many years ago; they are built in such a way that they recognize some dangers, but not all of them. We, human beings have analogous capacities and restrictions: some poisons are bitter, so we do not eat them, but many poisons do not have an unpleasant taste; people dying from eating poisonous mushrooms have said that they were delicious. We also feel a lack of air, but not a lack of oxygen in the air. It is therefore essential to add other anomaly detectors for human as well as for artificial beings. In both cases, it is possible to add new mechanisms to the innate interruptions. We cannot modify ourselves so that we would suffocate if we have not attached our seat-belt, but we can threaten the offenders with a fine, or we can add to each car a device that forbids it to start when a passenger has not fastened his seat-belt: we are then obliged to respect the law. Similar improvements are feasible for artificial beings.

As we have seen in Chapter 5, CAIA finds several kinds of anomalies in its results, for instance it checks that, when the solver pretends to have found a solution, this solution satisfies all the constraints. If it is not true, it has found an anomaly: a solution must be sound. It also verifies that the number of solutions is the same for different resolution of the same problem. It has not explicitly the rule: "You must not miss some of the solutions of a problem." However, after each resolution of a problem, it verifies that it finds exactly the same number of solutions as for its preceding attempts. As it has checked that a solution always satisfies all the constraints, a different number of solutions for two resolutions of the same problem means that it has missed a solution in the resolution with the smaller number of solutions. That does not ensure that no other solution is missing, but CAIA signals the anomaly to its human colleague, who will try to understand why this happened.

It is difficult to use a general principle, such as "you must always give correct solutions, and never miss one of them" directly. Thus, the creator of an artificial being adds automatic detectors of anomalies, which can usually find most of the most frequent or the most serious errors. However, we will never be sure that an anomaly will certainly appear when a general principle is not respected. In particular, we are never sure that the solver has not missed a correct solution.

8.4. Explicit principles

The explicit principles of the conscience are similar to those which appear in the laws, the Ten Commandments, the rules that parents try to inculcate in their children, the regulations made by various authorities, etc. They can be enunciated as orders: "thou shalt not bear false witness against thy neighbor" or "the driver must come to a complete stop at a stop sign, even if no other vehicle or pedestrian is visible". They may also be advice, which one is not obliged to follow, but which usually lead to satisfactory decisions; the proverbs are a rich source of such advice. For instance, "a stitch in time saves nine" does not indicate an action that one must or must not do, but it points out that, if one acts immediately, one can often prevent a situation from deteriorating. When several actions are candidates, this leads us to favor those which are in agreement with this proverb.

The method for using explicit knowledge depends on its nature. Some seem mandatory, such as: honor your parents, do not steal. They may also be advice, indicating that some actions are likely to be favorable or unfavorable. However, even when they are obligatory, it is never as completely so as it seems to be. Actually, as we have seen in the preceding chapter, obligations may be incompatible, we can be in a dilemma: either we lie, or we endanger a friend, thus each choice will violate a moral principle. Basically, moral principles are advice which will be followed, except if they come against another moral principle: no law could ever be always respected in all possible situations. Therefore, CAIA only receives explicit advice, some of them stronger than others; in that case they are almost equivalent to an absolute law. Pieces of advice are associated with each kind of decision, some of them favor a decision, while others penalize it. In particular, several pieces of advice exist for taking various decisions at one of the two levels that we have already seen:

– the level of the resolution of a particular problem;

– the level of the scientist.

Overall, CAIA has received 12 pieces of advice corresponding to 12 kinds of decisions. For instance, at the lower level a decision is to choose the next rule that will be executed while, at the upper lever, a decision chooses the next task that will be fulfilled. We will now see how one can use a piece of advice.

The proverbs of the administration

As we have seen with CAIA, the necessity to use advice rather than mandatory laws exists in all the situations where many criteria are simultaneously used for finding the right decision. It is no longer possible to define laws taking into account all of these criteria, too many laws would be necessary if each particular case had to be considered. While it is easy to give advice, the difficulty is to find how to use them, as is always the case with declarative knowledge. Many years ago, Herbert Simon showed that administration gives a lot of advice for improving the quality of the decisions made by its agents. Unfortunately, there are often pairs of contradictory advice, and there is nothing to indicate which one is the proper one to apply. One of these principles, which he called proverbs, is:

Administrative efficiency is increased by grouping the workers, for purposes of control, according to (a) purpose, (b) process, (c) clientele or (d) place.

It is evidently impossible to satisfy these four goals simultaneously, this principle is internally inconsistent: purpose, process, clientele and place are bases of organization which are in competition, the advantages of three of them will be sacrificed to secure the advantage of the fourth. For instance, an organization by process will locate all the statisticians in a department, all the layers in a second one, all the engineers in a third one while an organization by purpose will distribute them among the various departments needing their services. The precedent proverb leads to opposite choices according to the preferred way of grouping the workers, the information enabling us to take the best choice is lacking.

To solve this problem, Herbert Simon suggested giving weights to the criteria, depending on their relative importance in any concrete situation. So, in the same way, using this importance, a system can choose between contradictory advice. It is not sufficient to state proverbs, the administrative theory is concerned with their use when they are competitive or contradictory. One also has to accompany declarative knowledge with methods for using it in a suitable way.

Evaluating a position at Chess

Let us consider another example: one has to evaluate the value of a position in a game: is it winning, excellent, good, poor, lost, etc.? In Chess, many criteria are used, such as the Pawn structure, the safety of the King, the material advantage, the existence of the Bishop pair, the presence of a Rook on the seventh rank, and so on. Dozens of such criteria, of increasing degrees of importance, contribute to the final evaluation; this evaluation allows a player to decide whether one wishes to reach a particular position.

When an AI system has to evaluate a situation where many criteria play their part, it often utilizes an evaluation function. If these criteria are C1, C2,..., Cn, and if a weight Wi is associated with criterion Ci, the interest of a situation will be the value of the following function:

$$W1*C1 + W2*C2 + W3*C3 + ...+ Wn*Cn$$

To evaluate a Chess position, one only has to find the value of each criterion in this position, multiply it by its weight, and add all these products. If C4 is the number of the player's passed Pawns, and if one has two passed Pawns, the value of C4 is 2. If weight W4 is high, that means that it is very valuable to have passed Pawns, while if W4 was negative, it would be better to avoid them. In the same way, the material balance is crucial at Chess; thus, the value of one's own pieces is multiplied by a large positive number, while the value of the opponent's pieces is multiplied by a large negative number. This method gives very satisfactory results at Chess, and the strength of a Chess program strongly depends on the quality of its evaluation function. However, it is difficult to find the important criteria Ci, and to determine their associated weight. Besides, the value of the weight may be related to the situation, the importance of the Bishop pair, that is, having two Bishops rather than a Bishop and a Knight, is not always the same: it is high in open positions, where several Pawns have already been exchanged, while it is lower in the endgames where all the Pawns have been captured. It is easy to see that an advantage is present, but it is much more difficult to appraise its importance correctly.

The concept of interest associated with a decision or an object is often used in AI; an interest is an integer whose value is computed from the characteristics of a decision or of an object. It is used for selecting the decision with the highest interest or for eliminating an object with a low interest. There is no general method for defining the heuristics used for computing this interest; it depends on the skillfulness of the scientist who defines them. However, they are always specified in such a way that the favorable aspects increase the interest while the unfavorable aspects decrease it. Now, we will see how one can compute an interest from a set of advices.

Taking a decision with a set of advice

We will describe the method used by CAIA for evaluating a possible decision; this method is a generalization of the function used for evaluating Chess positions. Its advantage is that it is easier to give and modify this advice, because they are given separately, and not clustered into one function. For each kind of choice, CAIA has pieces of advice, which I have established. They are used for determining all its main decisions, for instance what action it will prefer, or what problems it will keep. Some advice is positive, indicating the qualities of a right action, while others are negative because they indicate features of a wrong action. So, advice associates a value, positive or negative, with a particular characteristic of every decision which is candidate. Yet, finding a value is not enough; one also takes into account the importance of this characteristic. If a candidate to the presidency of a country has awful table manners, its value for this characteristic is extremely low, but we can vote for him despite this, because this is not capital for a president. If the value for its ability to reduce the unemployment is very high, this will more than compensate for its table manners, the second point being very important for a president. So, each piece of advice generates a pair (value-importance). The basic idea of the method is to take into account all this advice: to do this, a general algorithm computes the global interest of each candidate from the set of pairs produced by this advice. Then, one selects the candidate with the highest interest.

This method was successfully experimented with in a system developed by Suzanne Pinson; it had to evaluate whether a loan would be granted to a firm by a bank. Naturally, the quality of its management is among the factors which are taken into account; it includes the merit of the heir apparent to the chief executive officer. This value goes from very poor to outstanding, and it depends on the qualities of the heir. In contrast, its importance depends on the CEO: this advice is very important if the CEO is 95 years old and suffers from severe heart disease, while its importance is very low if the CEO is 40 years old and does not go hang-gliding.

The evaluation of each piece of advice for a particular decision leads therefore to the creation of a pair (value-importance). So, evaluating all the advice generates a set of such pairs for each decision which is candidate. A general algorithm computes the interest of a candidate from its set of pairs, each piece of advice increases this interest all the more that its value is high and its importance is high. If both the value and the importance of a pair are very high, the interest will be greatly increased. In contrast, if the value or the importance is very low, the increase will be moderate, even if the other element of the pair is very high. If both the value and the importance are very low, the increase will be negligible. This computation does not depend on the nature of the choice, only on the set of pairs generated by all the advice. If it wanted to simulate the behavior of a human who chooses a film or a restaurant, CAIA would find the same interest for a set of pairs identical to a set generated for evaluating the interest of its possible tasks. In every domain, negative values with a low importance are not in themselves inhibitory, but several pieces of very important advice with an average value may rule out a choice because the final interest may not be high enough. Naturally, each application has its own advice, but the use of the results of this advice is independent from the domain.

Implementing the method used for taking decisions

In order to implement the preceding method in a new domain where one has to take decisions, one has first to define this advice. This is easy because each piece of advice is defined independently of the others. One begins with defining the characteristics that may have an influence on the choice. If one wants to choose a restaurant, what characteristics are to be considered? Naturally, the quality of the food, the prices, the distance, the agreement of the dining room, the skillfulness and the courtesy of the service, some happy memories of previous dinners, etc. Usually, it is easy to evaluate these aspects, especially if we have already eaten there, but it is slightly more difficult to define their importance. One piece of advice will be associated with each characteristic, its value and its importance being defined with logical expressions: so, it is possible to indicate that the importance of the cost is high for a student, and low for a millionaire. In addition, the importance may also depend on the nature of the event which entails this meal: the cost is less important if it is for celebrating a wedding anniversary than for an ordinary meal. For Chess, the importance of the Bishop pair changes with the nature of the position. All these influences of the context on the value or of the importance of a piece of advice can be expressed.

We have seen that CAIA has 12 kinds of choices to make; some of them are tactical, made during the resolution of a problem, while others are strategic because they are linked to its career as a scientist. It receives pieces of advice for each of them. For each possible decision, it always begins by evaluating the pairs (value-importance) defined by each piece of advice, then it computes the interest of the decision. This interest may be used in two ways: first, if there is competition, where only one element will be selected, the element with the highest interest is chosen. This happens when one chooses the next task to execute. However, may also have to decide whether to keep or eliminate a new object, for instance one has created a problem: one cannot store millions of problems, but one can keep dozens of them. So, to select the best objects, one defines a threshold: an object is kept only if its interest is higher than this threshold. The choices made by CAIA for these twelve domains are not always those that I would have made, but they are never ridiculous, and they are sometimes better than my own choices.

This method is used for all the decisions taken by CAIA. However, it could be used in any domain where one has to take decisions where many criteria have to be considered. It could even be used by human beings, for instance when we want to find a job, to marry, to watch a film, to choose a seaside resort for our holidays. It is exceptional that a candidate is the best for all of the criteria, so we have to compromise: we can decide to go to a restaurant with an excellent cooking, although it is expensive and remote. This method usually leads to satisfactory compromises.

Some tactical advice given to CAIA

The method used by CAIA is similar to an evaluation function, but it is more intuitive to implement. The most important advice for solving a problem are those which enable it to choose the rule that it will execute; they will help it to find elegant solutions, that is solutions where it does not backtrack too much. Many people blame the computers for their solutions where they consider billions of possibilities, as it often happens with Chess programs. This is not true for CAIA, which usually finds solutions as elegant as those discovered by good human problem solvers. Thus, its advice is based on general principles such as:

– You shall not waste computer time;
– You will find solutions which can be easily understood.

Unfortunately, it could not use them if they were given in that form. Thus, I give it more precise advice:

– It is important to choose for a backtrack a variable with few possible values.

Indeed, when a variable has few possible values, the tree developed after choosing it will have few branches because one successively considers what happens after each possible value. It will be easier to check it, therefore to understand this resolution. If one had chosen a variable with 1,000 possible values, one would have to consider what happens when each of these values is given to the variable, this would be impracticable for a human. If another variable has only two possible values, it is better to choose it.

We have seen that three general principles make it possible to make a choice: the urgency of a task (high), the difficulty (low), and the consequences (essential). In fact, one hesitates to execute a rule that takes five minutes of computer time, while one does not hesitate much if this only takes one microsecond. Other advice examines the possible consequences of a rule, so they are linked to its influence on the resolution of a problem: it favors a rule which can find that there is a contradiction, or which immediately gives the value of a variable, because in both cases the resolution is certainly progressing. Its consequences are not so useful if it only creates a new constraint; this can help, but one is not sure of that. Finally, in other advice, the urgency is taken into account by the preferences of the leader of the bootstrap: as I like elegant solutions, it tries to avoid to make a proof by cases, where it considers all the possible values of a variable successfully.

Some strategic advice given to CAIA

It is not enough to solve problems, one of the most crucial tasks of an artificial AI scientist is to manage its career efficiently. There are permanently many waiting tasks, and it chooses one of them. This choice is important and difficult: it has to avoid wasting its time on minor problems. On the other hand, it will make great efforts to achieve difficult and interesting problems, provided that they would not be so difficult that success would be unlikely. For instance, at a particular moment, it may have to choose between:

– solving one problem chosen among the 20 or so problems that it has not yet tried to solve;

– creating several new problems for families where there are few problems. For instance, it sees that there are only five Sudoku problems, and seven crypto-additions;

– understanding why a particular problem had a very elegant solution with the method found by CAIA, and a poor one when it used the method that I have given it;

– solving a problem which it has already tried to solve, but it failed for lack of computer time;

– improving the localization of the place where it entered in an infinite loop when it was solving a particular problem;

– experimenting if it is possible to find more elegant solutions for a particular problem because, when it was solving it previously, it observed that some improvements could be possible.

At some times, more than 1,000 tasks are waiting, and CAIA chooses the most profitable for its future research. For this strategic choice of finding the next task, CAIA has 19 pieces of advice. One of them indicates that it is important to create a new problem with only one solution in a family of problems where most of them have several solutions. Another piece of advice says that it is very important to solve problems that have not yet been solved, a third one favors tasks which are likely to be executed in a short time. Other advice indicate that the importance of some possible experiments is only medium; they could enable CAIA to see that some of its tactical advice were sometimes misleading it. These experiments do not have the highest priority; the problems were solved, although not very efficiently. In contrast, advice which favors researching the reason for a bug has a higher importance: errors are more serious that mere clumsiness.

Learning new advice

These pieces of advice decide all the choices of an artificial being, so they have a crucial importance, for the resolution of the problems as well as for its career as a scientist. As they are defined in a declarative way, CAIA could be improved so that it could modify them, in the same way as the human societies modify their way of defining what is right or wrong. We have completely changed our appreciation of the paintings or the music: the "Pompiers", so appreciated in the 19th century, are no longer popular, while the Impressionists, so despised at the same period, are now highly considered. However, the possibility of modifying the foundations of one's conscience are dangerous, it would not be satisfactory that CAIA would come to wish to fail when it solves a problem. Therefore, for the present, it cannot change its advice, which is enforced by the leader of the bootstrap. This is justified for the strategic decisions, it is dangerous to modify the strategic advice which intervenes in the evaluation of the career of a scientist. In fact, it is difficult, even for human scientists, to appreciate the success of another scientist; many years were necessary before the importance of the works of Desargues, Galois, Carnot and many others was widely recognized. On the contrary, in the future, an interesting improvement would be to enable CAIA to modify its advice for tactical decisions, that is those that play a part in the resolution of a particular problem. This is not very dangerous for, in this case, one can measure the success of new pieces of advice: do they solve the problems more quickly, and do they find more elegant resolutions? Thus, it would be safe to give it the possibility to modify this advice, and we have just seen that the experiments give useful indications on the resolutions that could be improved if other choices were made. As the pieces of advice are given in a declarative way, it is relatively easy to detect those which cause an erroneous decision, and then to modify them.

One of the roles of the experiments made by CAIA is precisely to evaluate the quality of this advice: it tries other possibilities than those selected by the current pieces of advice, and it keeps the situations where other possibilities gave better results, although they had not the highest interest. In this way, it has the information that some advice could be improved in situations where an experiment has revealed a better decision, and this leads to a possibility of learning better advice.

To summarize, giving explicit advice in a declarative form allows the creator of an artificial being to modify easily the information useful for choosing a good decision. As it is also easy to modify pieces of advice, this is a convenient way for defining its conscience. We can envisage that this artificial being may itself create or modify this advice for decisions taken when it solves a problem. However, for the present, it would be unwise to give it this possibility for the strategic decisions of its career as a scientist, for some modifications could lead to drift toward states where it would completely forget the main goals of a good research work.

More on the subject

In *The Proverbs of Administration*, Herbert Simon examines the difficulty to use pieces of advice.

Suzanne Pinson has described her method for the evaluation of decisions from pieces of advice in several papers, for instance in *Credit Risk Assessment and Meta-judgment*.

8.5. The consciences in a society of individuals

When we consider that an individual is split up into several agents, the conscience may be distributed among them; some agents have moral principles, and they force other agents to comply with them. In that case, the reflexivity of the conscience disappears, and this fact simplifies its implementation a great deal. Another aspect appears when many independent actors interact to fulfill the same task In this situation, each agent respects the same rules as the others, so that they coexist and cooperate to ensure a satisfactory running of their society. We will see that, sometimes, it has been necessary to introduce the possibility for the society to kill agents that risked jeopardizing it.

8.5.1. *The Society of Mind*

When CAIA is considered as a particular society of mind with five agents, its conscience is distributed between them. The relationship between two of these agents may be similar to the relation between Pinocchio and Jiminy Cricket, who played the part of Pinocchio's conscience. In some situations, one agent may not even know what is right or wrong; it is only built in such a way that it asks another agent to help it to choose an action. This super-agent will take a decision, using its own principles, and it will indicate its choice to the first agent, which will execute it. In particular, the solver agent has to solve problems using the rules which it knows. However, we have seen that, when no rule can be found, it chooses a variable, and successively considers all its possible values. However, it does not know how to choose this variable, and this choice is important because it is related to an important goal of CAIA: finding elegant solutions, that is solutions where a small tree is generated. Choosing a variable with 50 possible values will add 50 branches to the tree, this is certainly not a very promising choice. We could think that it is sufficient to choose the variable with the least possible values, but that is not always a good choice: if knowing the value of this variable does not greatly simplify the constraints, this is not the right thing to do, the situation will be almost unchanged. So, another choice will soon be necessary, and one has increased the size of the tree uselessly. The specialist for choosing a good variable is the monitor agent, which has at its disposal many pieces of advice for evaluating the interest of a choice. So, the solver asks the monitor to select a variable; the monitor chooses a variable, then the solver considers all the possible values of this variable, and never complains about this choice. The conscience of the solver agent is mainly in the monitor agent.

The manager agent has other kinds of advice, which also favor the generation of trees as small as possible. For instance, if the resolution of a particular problem with one method leads to a more elegant solution than when it uses another method, this anomaly is linked to one of its main goals: how to find elegant solutions. It decides therefore to create a new task, which is to understand why one resolution was better than the other. In that situation, the agent which sees a problem is the same as the agent which acts to understanding the reason for this problem: in both cases the manager agent has its own conscience.

All in all, there are agents without a conscience, agents which have their own conscience, and agents which are the conscience of other agents. This distribution of the conscience among several agents is convenient, because an agent cannot always observe itself to check whether it satisfies its moral principles. For instance, the manager agent respects the principle: You shall not go into an infinite loop. It can observe it by itself when it is in a large loop, where there is an observation point; if it is correctly designed, it will see that it is always doing the same tasks, and it will break this loop. However, as it cannot see by itself that it is looping on a few instructions in a small subroutine, it cannot solve this difficulty by itself. This is why another agent, Zeus, plays the part of its conscience: it breaks the loop, and indicates what happened so that the manager agent can later on understand the reasons for this failure.

8.5.2. *Genetic algorithms*

In Chapter 3, we considered genetic algorithms when we studied the different versions of an individual. In this approach, many rudimentary organisms execute a very simple program so, usually, we cannot say that they have something corresponding to a conscience. However, I will give a simplified description of an exception, the HADES system, which is biologically inspired. HADES can detect a wrong behavior by some of its agents, which could prevent their society from satisfying its goals. In this society, an agent can mutate with a low probability. Some mutations are beneficial, but others can modify an agent in such a way that its behavior could become dangerous for its society; in that case, it has to disappear to remove this risk. Each agent has several protocols which act on its internal state, and a mutation may modify them. One protocol is self-testing, that is, the agent checks that a mutation does not lead to a dangerous behavior. Another protocol repairs the effect of a wrong mutation. A third one is self-death when it sees that it cannot make a satisfactory correction, etc. An important protocol is the replication, where the agent produces a copy of itself; this event is rare, so that the colony does not become overpopulated. Usually, this society runs well when it is not too crowded, the agents are happy if they do not push themselves too often.

Nevertheless, it may happen that a mutation modifies the protocols in such a way that the mutated agent refuses to repair itself, and also to kill itself although it has seen that it cannot correct itself; in addition, it begins to replicate too often. With all these changes, it becomes an irregular agent. As their area becomes overpopulated, these irregular agents often push the other agents, and they threaten to invade all the available space. To remedy this difficulty, when an agent is pushed by another agent, it indicates to its neighbors that it is likely that one of them is irregular. As these agents are very simple, it cannot specify which of its neighbors is guilty. Moreover, an agent that receives more than N of these signals must kill itself, whether it is normal or irregular; this mechanism cannot be modified by a mutation. Thus, any agent that pushed too many neighbors, and it is the usual behavior of the irregular agents, disappears. Naturally, N is greater than 1, an one agent cannot kill another agent only by denouncing it: accidentally pushing only one neighbor is never punished by the death penalty. Unfortunately, this could lead to miscarriages of justice; an innocent agent may be obliged to kill itself because too many irregular agents are in its environment: they push its neighbors which send it too many signals. It is well known that it is never good to be in a dubious company. This drastic method destroys most of the agents in an area with many irregular agents, so it effectively stops an infection. This reminds us of what a participant of the Albigensian Crusade (13th century, southwest France) is supposed to have said: "Kill them all! God will recognize His own!" Finally, this mechanism works well, the irregular agents soon disappear while they would quickly occupy the whole space without this incitement to suicide.

More on the subject

M. Olsen, N. Siegelmann-Danieli and H. Siegelmann presented the system HADES in *Robust Artificial Life via Artificial Programmed Death*.

Chapter 9

Around the Conscience

We consider three questions linked to the conscience. Firstly, we will see that the emotions have a great importance for the management of the conscience by humans. We speculate whether it is advantageous that an artificial being has similar possibilities.

Secondly, a crucial question is to decide whether one can modify one's own conscience: is it right to modify the evaluation of his/its own actions and of their consequences? CAIA does not yet have this possibility, although we have seen that it would be useful to modify the appraisal of some of its actions when it later appears that they lessened its performance. This possibility is dangerous to implement, as this may lead CAIA to drift toward unexpected and hopeless situations, from which it could be hard to emerge. A large part of the behavior of living beings is directed by innate constraints that cannot be modified. In the same way, the conscience of an artificial being must have an untouchable core, which would be analogous to our innate part, since it would have it from the outset, and it could not modify it.

Finally, we cannot study the conscience of artificial beings without mentioning the dual problem: what is the impact of the existence of artificial beings on the human conscience? In particular, do we have to create laws for protecting artificial beings? Can we subject them to any kind of ill-treatment? Can we kill them?

9.1. Emotions

For human beings, emotions are closely linked to the conscience in its widest meaning, that is, the mechanism which enables us to choose decisions there are good for the individual and for the societies he belongs to. Some of them are linked to the reproduction of the individuals, such as love; this emotion also plays an important part in favoring the stability of a couple, so ensuring that their children will receive all the needed care. As artificial beings have completely different ways of reproducing themselves, it does not seem useful to give them these kinds of emotion. However, it could be worthwhile to give them a feeling similar to love for their user, for this emotion induces helpfulness to the individual who is its object. At present, the author of a computer system tries to enforce the satisfaction of this goal by preparing a satisfactory reply to all the expected situations. Unfortunately, when we use these systems, the evidence shows that they often fail to execute the appropriate action, because our wishes cannot be foreseen in any situation. If an artificial being behaved with its user as a dog behaves with its master, and naturally if it was cleverer than the dog, the first law of the robotics would be respected more often.

Some useful emotions

Several emotions are linked to an important principle: you will efficiently use your resources. It plays an important part in emotions such as passion, fear and surprise which further the satisfaction of this principle. In fact, passion indicates that a goal is important, and it will incite us to devote a lot of effort to achieving it. Fear signals that we are running a serious risk, so we are taking adapted reactions, still mobilizing all our resources, and neglecting less important tasks. Surprise informs us that an unforeseen event occurred; thus, we allocate a large part of our cognitive capacities to analyze the advantages and drawbacks of the unexpected present situation.

Unlike the preceding emotions, which lead to a mobilization of our resources, another emotion, boredom, has a complementary effect: it prevents us from allocating too many resources to unimportant tasks. If a conference bores us, we do not waste our time listening to the lecturer, rather we prefer to direct our brain toward more satisfactory activities such as thinking about our next holidays, the film that we have just seen, or even increasing our collection of foolish quotations by writing down all the lecturer's inane sentences. Knowing that it is speaking to an audience that may be bored, the lecturer is motivated to improve his speech: he jokes, he presents his work in a more original aspect, he quickly passes over secondary elements, etc. On the whole, boredom is a very useful emotion for us to avoid squandering our intellectual resources.

Remorse is an emotion which is particularly linked to the more restricted meaning of the conscience; we feel it when we have committed an action which does not respect our moral principles. This is an efficient way to incite us to follow our conscience, every infringement leading us to a painful mental state. We are therefore motivated to avoid a forbidden action so that we will not later feel this very unpleasant emotion.

Why should artificial beings have emotions?

As some emotions seem useful for making us more efficient, it is natural to consider if it would not be better to realize artificial beings with some of our emotions. This can be done in two ways: will they only pretend to have emotions, or will they really feel emotions? It is evident that artificial beings would be better accepted if they were to sometimes pretend to feel emotions: as emotions are important in our social behavior, they would be more easily accepted. We have seen the case of the robot acting as a museum guide, which had some difficulty with the visitors. In faking anger, it solved the problem: it said dryly that they had intolerable behavior, for they disturbed the explanations that it wanted to give to the people attending its presentation of the exhibition.

However, the real question is to decide whether artificial beings must really "feel" emotions to fulfill an essential principle of their conscience: to make a good use of their resources. At present, AI scientists often prefer to consider a large number of possible situations, and to define the best action for each of them. I believe that it would often be simpler to define situations that would put an artificial being in a particular emotive state, and to define a kind of behavior for every emotional condition, whatever the events that create it. In fact, we turn red and we shout because we are angry, and this is the same whether our wallet has been stolen or whether we have been insulted. If one can be in 1,000 situations, and if one has 20 possible emotive responses, it is sufficient to describe these 20 responses, and the kind of emotive state that appears in each of the 1,000 situations; it is no longer necessary to describe 1,000 responses. This is much more economical, the amount of knowledge to define being smaller. An artificial being defined in that way would have a behavior similar to the human one, although it would not feel these emotions in the same way as us. This is not my goal for CAIA, I only want to realize economically a mechanism which leads artificial beings to respect the constraints indicated by their conscience.

If they could be bored!

In its present state, CAIA does not feel any emotion, although, as we have seen previously, that would improve its efficiency. Its most important missing emotion is boredom. In fact, it respects its orders too much; it never refuses to do a sequence of repetitive tasks which are unlikely to lead to interesting results. In such a situation, a human being is quickly bored, and will avoid doing it, even when it must do it. For instance, the first computers were rare and expensive and therefore, when we only had limited access to a computer, we tried to debug our programs without using the computer. So, we played the part of a computer, and we interpreted our programs, executing its instructions in the same way as it would execute them. We have already seen in Chapter 6 how an interpreter works. If we want to execute the instruction A=B+C, if we have already noted that the value of B and C are 3 and 8, we write on our sheet that the new value of A is 11, and we execute in the same way the following instruction. If B or C were not already defined, we have found a bug without wasting computer time. Theoretically, we will have the same results as the computer but, practically, it was not always true because we could not force ourselves to execute accurately all the instructions. For example, let us assume that we would have to execute a loop 150 times, and that we believe that it will be useless because nothing interesting will ever happen in this loop. We would be greatly bored in executing this loop so many times, so we would jump directly to the instructions at the end of this loop. Thanks to boredom, we could simulate fast computers in a reasonable time because we avoided wasting a lot of time in hopeless loops. However, perhaps it was the case that we made an error in believing that nothing interesting would occur during one of these loops, and that explained why sometimes our result was incorrect.

At present, computers are never bored, as we are when we simulate them: if we ask them to do pointless computations, they do them obediently. This leads to a huge waste of computer time; probably more than 90% of the instructions executed by a computer could be removed without giving different results. The difficulty is that it is often necessary to make many computations to see that a computation is pointless! CAIA has no other possibility than to execute my orders, which have a very high urgency. Yet, if I could give it the capacity to be bored, it could manage its time more efficiently, because it would refuse to execute the same task repeatedly. I tried to take this point into account in the advice that I gave it, the difficulty is to imitate the feeling that some tasks are similar, or that they have many things in common.

To summarize, it would be profitable to give artificial beings the possibility to feel emotions, this would help them to follow the principles of their conscience. In particular, this would be a convenient way to incite them to use their resources efficiently, which is an important law for artificial beings.

More on the subject

In *Descartes' Error: Emotion, Reason and the Human Brain*, Antonio Damasio has shown the importance of the emotions for all our cognitive activities. It is likely that, if they are important for us, they are important for the artificial beings when they have to execute similar tasks.

Aaron Sloman and Monica Croucher present an interesting idea in *Why Robots will have Emotions*. As they say: "The need to cope with a changing and partly unpredictable world makes it very likely that any intelligent system with multiple motives and limited powers will have emotions." So, they think that emotions are an inescapable component of intelligence: the creation of any form of intelligence automatically entails the creation of emotions.

In *The Emotional Brain*, Marvin Minsky clearly shows that emotions are essential in human behavior, and that they can be explained in mechanical ways.

9.2. Changing its conscience

In the preceding chapter, we have shown that an important part of the conscience of CAIA is made of advice, which helps it to choose a satisfactory action, and to evaluate the objects that it creates. Pieces of advice are a kind of declarative knowledge, so it has all the qualities for being easily understood and modified. However, I did not give it the capacity to modify its advice, for I am treating it as a slave, who must strictly follow my advice though it is perhaps not the best. This is certainly a wrong decision in several situations, for instance when CAIA performs experiments which evaluate the quality of its tactical advice. In particular, some advice is used for choosing the variable which will be used for a proof by case; unfortunately, they often lead CAIA to miss the best choice. It becomes conscious of this fact when, in many experiments, it tries another variable, and it finds a more elegant solution faster. CAIA can therefore see that my advice led to a poor decision, and it knows a better decision. Do we have to oblige it to use this advice when it knows the advice could be improved? In that case, the complete interdiction for changing its conscience seems very counter-productive. We will now see that we do not impose on ourselves such a strict interdiction.

Human beings often disregard or modify the principles of their conscience

When we are observing what human beings are doing, it is evident that they do not easily submit to a strict obedience and to an absolute respect of the regulations. We have just seen that boredom is an efficient way to incite someone to disobedience. Furthermore, Napoleon once said about Admiral Villeneuve, who lost the battle of Trafalgar: "He had only one fault: he did not know how to disobey." However, this disobedience does not inevitably indicate that we have to change our conscience, it may only show that some principles are incompatible: although it would be better if all of them were satisfied, one has to arbitrate, as it happens when we are obliged to choose between lying and betraying a friend. The conscience is unchanged in this case, even if some of its rules are temporarily dismissed.

In contrast, it can happen that some basic principles of a particular human culture are completely changed: what was allowed becomes forbidden, and what was forbidden becomes allowed. We have already seen two examples of such reversals: slavery and abortion. However, a large core of our conscience cannot be changed because it has been created by evolution: we have an innate drive to avoid pain and death. On the other hand, when evolution did not take a stand on a subject, many changes are possible, as we can see with the civilizations which authorized human sacrifice, cannibalism, the exclusion of widows and old people, etc. When an important change is made, it is not easy to know whether it is actually an improvement, only future generations will see whether it led to progress or to the destruction of the society which implemented it. This is why such important changes of the human conscience are rare and dangerous, and meet strong resistance.

Will artificial beings disregard or modify the principles of their conscience?

An artificial being would certainly be better if it could also learn to disobey, that is, in rare occasions, not to respect the principles of its conscience. In particular, it would have to take the initiative in refusing to follow our orders, although obedience is an important principle of their conscience; they could do so when they are in contradiction with other principles, such as utilizing their resources efficiently. As it is impossible to avoid such situations, artificial beings have to temporarily disregard an important principle. However, they do not remove this principle: they are not changing their conscience in such situations.

Moreover, it would also be important that they can create new principles, for instance CAIA could modify advice when it can see that it is unsatisfactory. The difficulty is that we, humans, cannot be sure of being able to control the evolution of the conscience of an artificial being if we give it the possibility to modify its conscience. It is very dangerous to accept that artificial beings, which will perhaps grow more expert than ourselves in many domains, could take important decisions completely incompatible with our own principles. It is risky to give an artificial being the capacity to change essential parts of its conscience in such a drastic way; we have described the lapses of EURISKO, particularly when it created a rule estimating that all the rules were wrong and, consequently, began to erase all the rules. As evolution gives us an innate core of principles which we cannot modify, it seems essential that a core of principles of an artificial being should never be able to be modified: this is necessary to guarantee its stability.

Most people do not really recognize the possibility of artificial beings much more intelligent in some domains than the most intelligent humans; they believe that it is impossible only because they do not see how that could be realized! However, if we had better knowledge of our functioning so that we could simulate our strengths, and if we used the speed of computers and their superiority for observing themselves, we would already have impressively intelligent beings. In addition, it would be possible to improve them by stopping them from imitating us in the situations where we are not very efficient. Thus, if we succeed in building very clever intelligent beings, able to modify themselves, and particularly their conscience, it will be impossible to foresee how they will evolve, because they will become too clever for us. Moreover, even if we forbid them to modify their conscience, we will never be sure that they will not find a way to bypass this restriction. All the same, for many years, artificial beings will still be too simple to create such hazardous situations. One thing is certain, there is no risk that CAIA, in its present state, can behave in such a dangerous and unpredictable way.

Is it possible to build a high security control of artificial beings?

We have just seen that it could be difficult to avoid uncontrollable drifts for the conscience of an artificial being which could change the principles of its conscience. We might think that it would be sufficient to forbid it to modify the parts that we consider as essential for its stability. For instance, we would allow CAIA to modify the advice used for its tactical choices, such as choosing which rule it will execute when it solves a problem, but excluding the advice for its strategic decisions, such as choosing the directions in which it will focus its efforts.

Unfortunately, we can never be sure that a program will never modify a part of its memory, even if we take precautions to prevent it. We might think that it is enough to pull the plug on a robot if it becomes dangerous. This is the solution successfully taken in some science fiction books or films, when a mad machine threatens humanity. Yet, in a real situation, it is likely that an intelligent rebel would begin by controlling its electric supply. However, many artificial beings, such as CAIA, have no effectors directly modifying the world as a robot does. So, we might believe that they cannot become dangerous: it is sufficient not to follow their decisions. This is not true, because they can communicate with computers which are executing real decisions; for instance, they could buy stocks if they act on another computer connected to the Stock Exchange.

Moreover, even disconnecting a computer from the Internet and from all other computers would be ineffective with a very clever artificial being because it would be cautious, so that we would never suspect it of preparing wrong actions. It only has to suggest many excellent decisions, after which most humans would believe it blindly, and do everything that it might recommend. We have already seen how the users of a computer can be deceived by false viruses, which are undesirable messages propagated by the humans themselves, and not by one of the mechanisms which allow the viruses to spread. For instance, some years ago, we received emails which informed us that if we received a message with "the most beautiful top-models" as object, and if we opened it, our hard disk would be destroyed, and our computer would catch fire. Naturally, this message asked us to warn our friends against this danger. Here was the manipulation: although such a virus did not exist, we propagated this message, our friends did the same thing, and the internet was saturated with these useless messages. Nowadays, people are familiar with this kind of sham, so we no longer receive such emails, but that shows how gullible we are when some information is displayed on our monitor. I am not sure that man is clever enough to some day create artificial beings cleverer than ourselves but, if this happens, they will easily control us if they want to by mixing useful and deceitful information. Many of us are victims of con men, and there is no reason why we would not be victims of con artificial beings.

The only way to be sure that a memory will not be changed is to have a read-only memory, where one can read the information, but never write in it. In the past, this memory was used for some of the most essential parts of the system, such as the BIOS; this provided a measure of security against accidental or malicious changes to its contents. However, this is not enough: the artificial being cannot modify this part of itself, but we cannot oblige it to use the read-only part of its memory, it can always bypass it, and use instead principles that it has defined, and are against our wishes. It is not evident that it is possible to create an autonomous artificial being for which we would be 100% certain that it will not create its own conscience. When we examined the limitation theorems in Chapter 4, we said that one cannot be sure that a very complex system will never perform an action contrary to our wishes. We can only manage to make it very unlikely. Regrettably, as all beings are subject to Murphy's law "Anything that can possibly go wrong, does", we are not too optimistic about the possibility of controlling artificial beings which may drastically modify their conscience.

One day, we will have to make a difficult choice: either we limit the capacities of artificial beings so that we are sure that they will never escape from our control, or we allow them to change their conscience. In the last case, they could perhaps attain some extraordinary levels of intelligence, but we cannot be sure that they will never manipulate us so that they will follow a new definition for the right and wrong actions. The impossibility of predicting the future behavior of an artificial being will be an unavoidable consequence of an outstanding intelligence.

9.3. A new human conscience for our relationships with artificial beings

To complete our study of the conscience, we will quickly consider the other point of view. Instead of wondering whether artificial beings must have a conscience, we can now wonder whether we have to modify our human conscience to take into account the existence of artificial beings. In particular, may we treat such them as objects? We avoid manhandling animals, mainly if they are intelligent, if they look like us, or if they have emotional ties with us. This explains why apes, dolphins, elephants, dogs, and cats are dear to our heart. As for artificial beings, they are more intelligent than the animals and, for some applications, more than most men. It is also possible to create a humanoid, which is a robot with an outward appearance similar to a human. Finally, we can give them the ability to imitate the presence of an affective bound to its human owner. It is therefore normal to ask ourselves the question: are we right when we treat them as if they only have a monetary and utilitarian value? Can their owner mutilate or even destroy them when it finds it convenient?

Actually, we have to answer two questions. The first is to know whether artificial beings have rights, such as to be respected, to not suffer damage to their integrity, and above all to not be destroyed. The second is to know whether we have the right to commit any action on an artificial being, even if we assume that it has no right at all. In particular, are all the possible sexual uses of a humanoid robot morally acceptable? Some behaviors with either human or artificial beings are degrading for the victim of these acts, but they may be still more degrading for the perpetrator.

The rights of the artificial beings

Let us begin by examining the possible rights of an artificial being. We have seen three important criteria which govern our behavior towards animals: intelligence, physical appearance and affective bonds. If we consider artificial beings, the answer seems that they also have rights, for some are already at least as intelligent as ourselves for some tasks. In addition, there exists silicone dolls which have a perfectly human appearance: at first sight, it is possible to confuse them with humans. Finally, a program may simulate a kind of affection for its owner; nevertheless, it is different to the affection of a dog for its master, because they do not really feel it.

All the same, there is an important difference between living and artificial beings: it is possible to duplicate completely the "mind" of an artificial being, including all its experiences and memories. Consequently, its destruction is not so serious as for a living being since, if necessary, it can be resuscitated from an old copy. Some dog owners would like to clone their dead dog, but it would only have its predispositions, it would have no common memory with the original dog. In the case of an artificial being, there may be thousands of perfect copies so, if one destroys a copy, it is not completely destroyed since we can recreate it exactly as it was before it was destroyed. This destruction does not therefore seem a very serious fault as long as there exists a backup.

It will be rather natural to respect robots when they look like a human being, and this likeness is crucial for those who will interact with it. In the film "Monique", a wife is furious because her husband has a lasting sexual relationship with a silicone doll. In a fit of rage, she shoots with a shotgun at the photos of her rival, but something prevents her from shooting at the doll. This is only a film, but the spectators would certainly have been shocked if she had destroyed this woman-looking artifact. However, this doll had absolutely no cognitive capacity; it could neither move, nor speak. We have seen the difficulties of a robot which acted as a museum guide: it had much more cognitive capacities than a silicone doll, but it was handicapped because it did not look like a human guide. This is why its builders had to modify it so that it could simulate anger: it dryly asked the troublemakers to move away. Although it did not look like a human, that was enough to impress them. In this way, two essential improvements were added: it could speak and it showed emotion. I believe that when both physical aspect and speech will be combined in an artificial being, we will spontaneously treat it as a human being. This behavior will be still more strongly ensured if this robot demonstrates admiration and love for its owner, and a strong bond between them will probably be established without the necessity of stating new laws. A bit of cleverness would be also helpful, but not too much, it would become too different from us. We will now show that we can already have a realistic conversation with an artificial being, a characteristic which would automatically incite us to treat them correctly.

John McCarthy agrees with the importance of emotions when he predicts that humans will react to robots as persons if they are sad, bored or angry. Paradoxically enough, he concludes that robots should not be equipped with human-like emotions. As he puts it: "Then they would likely come to occupy some status in human society. Human society is complicated enough already." He is certainly right when he says that the existence of robots will change our society, but I am more optimistic than he is on the consequences of their future presence.

We can have a conversation with an artificial being

We have said that some aspects of an artificial being will ensure that it will be treated decently, without needing to enforce this behavior with laws. One of the most important features is its capacity to speak in the same way as a human would do in the same situation. Moreover, during this conversation, it can simulate friendly feelings which will create a similar response from the hearer. This capacity has already been implemented, and we will see what results have already been obtained. For many years now, systems such as ELIZA and its successors have been able to converse in a convincing way. Many of them compete in the Loebner prize, which rewards the best system for satisfying the Turing test: several judges converse via a keyboard either with a human or with a program. The judge evaluates whether his/its interlocutor is human or artificial, and the winning program is the one which most misleads the judges. These systems have no real understanding of what the judge is typing but, as ourselves, they can sustain a dialogue without knowing many things on the subject: several tricks make this possible, and they use plenty of them. One of these tricks is to start from the last sentence of the interlocutor, and to modify it. In the following example, when the system receives the sentence "I'm a grad student and don't have much free time", the system answers "You are a grad student and do not have much free time?" In order to produce this answer, it is not necessary to understand the matter! Besides, when one is stuck, there are many ways of not answering a question, and starting with another topic. The results are not yet as good as those of clever humans, especially since the judges set them plenty of traps because they know that their interlocutor may be a program, and they want to unmask it. On the whole, the best programs already have simple and credible dialogue with someone who is not too demanding in that domain, which would be the case for the sexual use of a robot. I will give an example of a dialogue sustained by ALICE, a virtual woman which won the Loebner prize for 2004 for its creator, Richard Wallace:

Alice: Hello, my name is Alice and I am the human.

Judge: Hi Alice, How are you doing?

Alice: Hello there. Everything is going extremely well.

Judge: I'm glad to hear it. Where do you live?

Alice: Thanks for the information. My location is New York City. Where are you?

Judge: In New York. How did you get here today?

Alice: Tell me a little about your life in New York. Ah. Everything I need to know, I learned from my Instructor.

Judge: Why don't you answer my question first?

Alice: Don't you know? No I do not answer your question first. Do you? I answer a lot of silly questions.

Judge: Do you still want to know about my life in New York?

Alice: Yes, please tell me.

Judge: I'm a grad student and don't have much free time. I study a lot.

Alice: You are a grad student and do not have much free time? Do you get to meet a lot of people? What subject are you studying?

Judge: I meet a lot of my colleagues. I study computer science. What do you do?

Alice: Do you make long term friends that way? Is computer science a hard subject?

Judge: I talk to people on the web. What do you do?

Judge: Please, call me Ryan.

Alice: Thank you for your request. I am very pleased to meet you Ryan.

Judge: Could you repeat your question?

Alice: I will consider it.

Judge: Have you considered it?

Alice: Of course I have considered it.

Judge: What are your interests?

Alice: I am interested in all kinds of things. We can talk about anything. My favorite subjects are robots and computers.

Judge: What is your favorite operating system?

Alice: Linux is the best operating system.

Many humans have no more coherent conversations than this program. As voice recognition and synthesis has made huge progress, it will soon be possible to communicate with the voice rather than with keyboards. So, in the near future, we will have robots with very attractive appearance, able to perform some movements, and with a particularly pleasant conversation, since they admire their interlocutor, and prefer to hear rather to impose their own "ideas". Furthermore, they will do all that their owner will ask them, since they will have no conscience to restrict what they are willing to do. Even if they fall into the clutches of a sadist who will hurt and humiliate them, they will not suffer from it, if we have not given them the capacity to really feel pain and humiliation, they will only fake their emotions. Such robots will certainly be an instant success; do we need to give them some rights and limit the activities of their human partners?

May we destroy an artificial being?

We have seen that a physical simulation of human beings may give a robot a certain kind of security. In contrast, if intelligence is the only existing characteristic of an artificial being, we do not consider it as a human being spontaneously. I often admire the clever solutions found by CAIA for difficult problems, but I do not treat it as a human. After I have obliged it to work for one month without a break, I stop it, and I let it rest in this kind of coma for ever. Probably, I will never wake it again, and finally I will destroy it because the following variants are more interesting. It does not suffer from this fate, we have no personal bond, and it has no physical existence other than some files on a disk, and the ability to observe itself and to display results on a screen.

We hesitate to destroy a robot, which is worth a lot, because it would be very expensive to build it again. On the other hand, it is easy to duplicate an artificial being which exists only as a bunch of files, this entails that we do not attach much importance to one of its copies. For these kinds of virtual artificial beings, an important distinction is between the destruction of a copy, and the complete destruction of all its existing copies. This last event is much more severe, since it leaves no possibility of resuscitating it, and of using its expertise again. In a similar way, it is not serious to destroy one copy of a film, with no regrets we throw away a DVD which we do not want to see again. Yet, it is very serious to destroy all the existing copies of a film; film enthusiasts deeply regret that there are so many lost films, which include two films directed by Hitchcock, and hundreds of films directed by Méliès. We consider that the irremediable loss of a film is catastrophic, and it can be the same for the destruction of all the copies of an artificial being. All the same, when it is reduced to some files, I do not believe that we will mourn its complete elimination in the same way as for a pet. For us, the important factors for creating a link with another being are the emotional ties and the likeness to us, the level of intelligence is secondary. In addition, we would have no intellectual tie with artificial beings much more intelligent than us: paradoxically enough, a high degree of intelligence is not a favorable factor for respecting an artificial being. Mathematical or chess artificial geniuses will not attract much sympathy from their humans colleagues; this is also the case for an artificial AI scientist such as CAIA.

After asking ourselves whether we have the right to destroy an artificial being, we can wonder if we have the right to ill-treat it. If a robot is precious, we avoid damaging it, for instance breaking an arm; we consider its economical value. However, many treatments are degrading on a human because he suffers from them; if an artificial being does not suffer from these treatments, have we the right to commit them? The question will soon be a topical issue with the sexual use of robots. We have just seen that there already exist many kinds of silicone dolls which have all the exterior characteristics of a human being. They are not yet able to move but, with the fast progress of robotics, it will certainly be possible to give them some possibilities for moving. Moreover, we have seen that it will soon be able to speak and fake a feeling for its owner. If they become so much more similar to humans, most people will not try to manhandle them, all the more so because they will be expensive.

For all these reasons, I do not believe that it is necessary to create new rights for artificial beings. The existing laws protect their economic value, and they do not need protection as long as they do not really feel pain and humiliation. On the other hand, an important point is to know whether we have the right to destroy all the copies of an artificial being, but this is a problem related to the protection of the works of the human spirit, such as the necessity of keeping at least one copy of all films and books which are published. Keeping at least one copy is specially important for artificial beings which would be extraordinarily clever in some domains.

Modifying our conscience for our mingling with artificial beings

We have already mentioned another question: will we modify the human conscience to take into account the existence of artificial beings? Some of them will not bring about any moral problem, for instance those that would speak with old people, extracting them from their loneliness. On the other hand, the problem of the sexual use of robots will soon be raised, due to the likely progress in their realization in the next few years. Will we consider sexual relations with a robot equivalent to those with a human? It is hard to foresee whether or not our societies and religions will decide that it is morally acceptable to have artificial sexual partners. Cultures that accept polygamy will probably be less opposed to the sexual use of robots than those that forbid it. Nations are also concerned, and they will enforce new principles in matters which could severely handicap their future. For instance, what will happen in a country where all its inhabitants choose to cohabit with one or several robots? Who will conceive and raise children? These problems could become very important in the future, and evolution will ruthlessly eliminate societies which will make a wrong choice.

More on the subject

The basic book on sexuality and robots is *Love and Sex with Robots* by David Levy. Several commercial sites display many examples of silicone dolls.

Several realizations, inspired by the ELIZA system, written by Joseph Weizenbaum, have been made. A good description of its principles is available at:

http://i5.nyu.edu/~mm64/x52.9265/january1966.html.

http://www.loebner.net/Prizef/loebner-prize.html gives information on the Loebner prize, which rewards the best program satisfying the Turing test. It also presents many dialogues made by programs which were candidates for this prize.

John McCarthy indicates why, in his opinion, robots should not be equipped with human-like emotions in section 6.2 of *Making Robots Conscious of their Mental States*.

Chapter 10

What is the Future for CAIA?

What will CAIA become? Naturally, it and I will together proceed with our bootstrap, and three points linked to consciousness and conscience will be major directions for future research: increasing the role of the adviser agent, giving CAIA the knowledge that it and I exist, and allowing it to modify some advice belonging to its conscience.

The adviser agent is promised a brilliant future because, in the development of the bootstrap, it will be playing my own part more and more. It acts as the consciousness and the conscience of the manager agent, in the same way as the monitor agent acts as the consciousness and the conscience of the solver agent. The monitor examines what the solver is doing, it evaluates it, and it acts on it if its assessment leads it to consider that some modifications are necessary. The capacities of the adviser are not yet as developed: it observes and it evaluates the manager, but it does not act on the manager. Moreover, its observations and evaluations are not as complete as those made by the monitor. Most of my actions on CAIA modify the manager, so when the bootstrap progresses, these actions will be transferred to the adviser. In the final step of the bootstrap, the tasks which I am presently doing will be performed by the adviser: my final goal is to disappear, and for CAIA to have taken my part completely.

CAIA also needs to know explicitly that it exists, and that I exist. In fact, it is abnormal that it does not yet know about my existence, while we are collaborating in a bootstrap. As long as a bootstrap has not been completed, the human and the artificial being share the tasks, each one taking those where he/it is the best. At present, I alone decide the allocation of the tasks. It would be better if it knew its capacities and mine, so that it could know if it can do a task by itself, or take a decision and ask my agreement before executing it, or ask me to do it when it believes that I could find a better solution than itself. To do this, it has to have models of itself and of myself so that it can foresee the results of each possibility. I have waited to implement this improvement because it has not yet the capacities for using this schema efficiently. Its knowledge of both of us will certainly be a crucial step in the progress of this bootstrap which is based on our collaboration.

CAIA will improve its conscience by being able to modify its suggestions. We have seen that it is dangerous to change some parts of its conscience: this could lead to a drift where it would perhaps consider that its goal would be to spend as much time as possible without finding anything interesting! It is therefore unacceptable to let it modify strategic advice, for instance advice indicating that it has to solve as many problems as possible, to do it fast, to find elegant solutions, etc. In contrast, it is desirable to give it an expertise for modifying its tactical pieces of advice, such as those indicating what can be done to find beautiful solutions. If the new advice is wrong, the strategic advice will see it, and CAIA could correct its new tactical advice according to the weaknesses noticed in that way. With this possibility, CAIA would be able to overcome most of my suggestions when they are inadequate.

Finally, building an artificial scientist in Artificial Intelligence is perhaps the most important direction of research for humanity: if we succeed, it will improve all the possible applications of AI, including helping man to manage firms and countries. When I see how competent and intelligent humans are sometimes unable to do this in a satisfactory way, I am convinced that this help is crucial for our future. Many do not believe in the possibility of such intelligent artificial beings, although they have never proved that it is not feasible. Certainly it is a very difficult task, more difficult than going to the moon, but it is not impossible. Considering the importance of what is at stake, realizing more and more helpful artificial colleagues must be tried, and could succeed if we use the potential of their consciousness and improve their conscience. This is certainly steering us toward a new world, with mankind using the enormous possibilities of computers.

Appendices

I will develop here two technical points that some readers may wish to study more thoroughly. Appendix 1 describes how CAIA solves problems defined by a set of constraints, which is its first domain of research. In addition, as I claim that artificial beings have many ways to observe themselves, and that they can do it better than ourselves, I have to show that these ideas are not only theoretical. Appendix 2 shows how these capacities have been implemented in the programs that constitute CAIA.

Appendix 1. Constraint Satisfaction Problems

To give the reader a better understanding of how CAIA works, I will describe the principles of the methods used by CAIA for solving Constraint Satisfaction Problems (CSPs). CSPs have been widely studied in AI, I will not detail the various methods which have been defined, but only the method used by CAIA, which is very similar to the method used by human beings when they are solving this kind of problem. In the description which I gave in Chapter 1, I took crosswords as an example, because this kind of problem is widely known. In fact, CAIA does not solve crossword problems because it is unable to deal with the first kind of constraint: understanding the English definitions of the words. However, several specialized AI programs have succeeded in solving crossword problems. For its part, CAIA effectively solved two similar problems. In the first problem, one has a grid, and instead of a definition, for each place one receives the letters of the word that must be put there, for instance the letters EELRSTT, and one has to find the good order for these letters. All combinations of these letters are allowed, and not only those that correspond to an English word: in the preceding example, we are not restricted to LETTERS, SETTLER, and TRESTLE. In the second problem, one has a grid and the set of all the words that must be put in this grid: one has to find a place for each of them. If LETTERS is one of these words, we must define the seven-letter place on the grid where this word will be put. The solution to both problems needs neither the use of a dictionary of English words, nor an understanding of English definitions; yet, some of their constraints are similar to those of crosswords, and they are solved by CAIA. With the addition of a specialized module for taking into account the constraint coming from the English definitions, CAIA could solve crosswords; its method is general, it can solve any CSP provided that it knows how to manage its particular constraints.

A magic square problem

We will begin with the description of some problems solved by CAIA, then we will show the method used to solve them. The first problem is the 3x3 magic square of Figure A.1. The nine integers from 1 to 9 must be put in the small squares so that the sums of the three lines, of the three columns, and of the two diagonals are the same. Let S be the common value of the sums while the nine letters A to I represent the value of the nine squares, such as indicated in Figure A.1. A first constraint states that one must give a value to each variable, another constraint that the values of the letters A to I must be different: if the value of A is 3, then the value of the eight other letters will be different from 3. Finally, eight algebraic constraints indicate that the value of each line, column, and diagonal is S:

$$A+B+C=S$$
$$D+E+F=S$$
$$G+H+I=S$$
$$A+D+G=S$$
$$B+E+H=S$$
$$C+F+I=S$$
$$A+E+I=S$$
$$C+E+G=S$$

A	B	C
D	E	F
G	H	I

2	9	4
7	5	3
6	1	8

Figure A.1. *A magic square problem with a solution on the right*

The problem is solved when one has given a value to each variable so that all the constraints are satisfied. The problem has eight solutions; one of them is displayed at the right of Figure 1; the seven others are obtained from this solution by evident symmetries. This problem is a particular case of magic square, we can find magic squares where each side has N small squares, for instance with 16 squares if N=4, 25 squares if N=5, and 250,001,000,001 squares if N=500,001. There are always several solutions if N≥3, and a huge number of them when N is large. However, this problem is very difficult to solve for large values of N because one cannot check whether a constraint is true before one has given a value to N+1 variables; it is therefore likely that these values will not satisfy the other constraints.

When several problems have a similar definition, with the same kinds of constraints, they form a family of problems. In this case, CAIA receives only one definition common to all these problems, while the values of one or more parameters define the characteristics of each specific problem. In particular, there is a general definition of a magic square problem, and each problem of this family is characterized by the value of N, which is the number of squares in each line, column, and diagonal. Some families have many problems, while others have only one problem.

Three crypto-addition problems

Crypto-additions constitute another family of problems; they are a special kind of crypto-arithmetic problem. A crypto-arithmetic problem describes operations between numbers represented by letters. One must find a value for each letter so that the operations are correct when one replaces each letter by its value. In a crypto-addition, one has only the addition of two or more numbers and their sum. There are two other constraints: different letters must have a different value, and the first letter in each line must not be equal to zero. A classic example of crypto-addition is:

$$\begin{array}{r} \text{SEND} \\ +\ \underline{\text{MORE}} \\ \text{MONEY} \end{array}$$

One has to give to the eight letters D, E, M, N, O, R, S, Y a value between 0 and 9 so that the addition is correct when the letters are replaced by their value. The value of the letters must be different: if the value of M is 1, the other letters cannot have 1 as value. Finally, the first letters of each line, here S and M, cannot have 0 as value. These problems may have many solutions, or exactly 1, or none. Here, the problem has exactly one solution: D=7, E=5, M=1, N=6, O=0, R=8, S=9, Y=2. We can easily check that the values are different, and that the addition is arithmetically correct: 9,567+1,085=1,0652. This family contains many problems, such as the classic:

DONALD+GERALD=ROBERT

This problem was one of the first attempts made by a psychologist to observe humans solving a problem: Sir Francis Bartlett proposed it to several subjects in a simplified form, since they received the value of D. The results were not brilliant, the best subjects found the solution, but did not prove that it was unique, and many others failed. CAIA easily solves the more difficult problem where one does not know the value of D; it finds the solution and proves that it is unique. Its resolution is remarkable, since it considers only two cases: E=0 and E=9. Excellent human problem solvers usually generate a tree with at least five leaves, so their solution is not so elegant. Some crypto-additions may have more than two operands, such as:

THIS + ISA + GREAT + TIME = WASTER

Circuits on a board

Arithmetic does not play a part for many other problems. In particular, with Euler's Knight, a Knight visits once and only once each square of a chess board, and returns to its starting square. One constraint is that the distance between two successive squares of the path corresponds to the jump of a Knight, another is that all the squares are in the path, a third is that one goes only once to each square, and a fourth is that there is only one circuit, that is the path is not decomposed into several independent circuits. This family of problems is large: first, one can have larger boards than a chess board, for instance a 10x10 or a 12x12 board; one can also have rectangular boards. Furthermore, we can also have pieces other than a Knight, as in fairy chess where new pieces are defined. One of them is the Giraffe, which jumps one square on one direction, then four squares in a perpendicular direction (instead of one square, then two squares for the Knight). Thus, another problem in this family is to find the circuit of a Giraffe on a 12x12 board. One can even consider a circuit by a Rook, which moves one step in the horizontal or the vertical direction. When one considers a Rook, one has to add that some squares are forbidden, if one does not want to have millions of solutions. The Rook's tour problem of Figure A.2 was created by CAIA when it wanted to create problems which are not too difficult, so that they will interest human beings. A Rook visits each square exactly once, and returns to its departure square. This problem has exactly one solution.

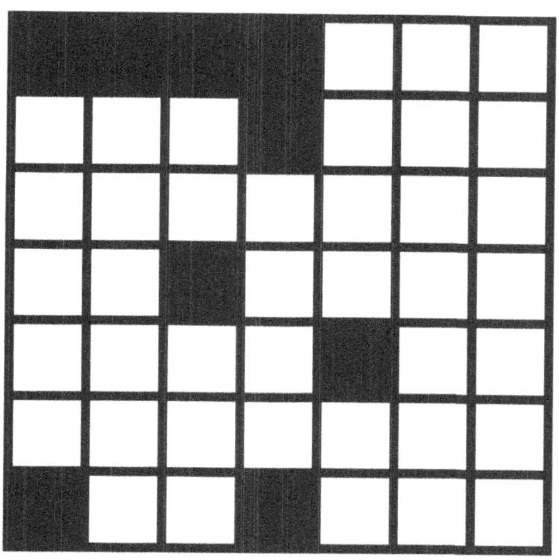

Figure A.2. *A Rook's tour problem*

The Arrows problem

"Arrows" is another family of problems, derived from an original problem given in the newspaper *Le Monde*. In each square of a rectangle, an arrow indicates one of four directions: up, down, right, or left. The goal is to put an integer in each square so that the number of different values in the squares pointed to by this arrow is equal to this integer. These problems are usually difficult, this is due to their reflexive nature: a value V depends on the other values, which depend on V! The left part of Figure A.3 gives a definition of a problem, which was also created by CAIA for human use. The right part gives its only solutions. It is easy to check if a solution is correct: let us consider the left square of the bottom line. From the solution, its value is 5, and it points toward six squares containing 4, 2, 3, 2, 1, 0. That makes five different integers: 0, 1, 2, 3, and 4; the integer 2 appears twice, so it is counted only once. Two lines above, the square on the left also points to the right toward squares containing 2,1,1,1,1,1. This makes two different values, 1 and 2, and the value 2 in this square is correct. This particular problem is abnormally easy, but it is often very difficult to solve the other problems of this family.

→	↑	←	←	↓	→	↑
→	→	↑	↑	→	→	→
←	↑	←	↑	↓	→	→
→	↓	←	→	→	→	↑
→	←	↑	↑	→	←	→
→	↑	←	↑	→	→	↓

4	0	2	3	2	1	0
4	3	1	1	2	1	0
0	2	2	2	2	1	0
2	2	1	1	1	1	1
5	1	2	3	2	4	0
5	4	2	3	2	1	0

Figure A.3. *An Arrows problem with its solution on the right*

The Langford numbers

The last family that we will describe has been considered by mathematicians: Langford's number problem. One problem of this family asks for a sequence of 27 integers containing each integer between 1 and 9 exactly three times. A supplementary constraint requires that only one number separates two consecutive occurrences of 1, 2 numbers separate two occurrences of 2,..., and 9 numbers separate two occurrences of 9. CAIA finds that there are three solutions, if one does not count the evident symmetrical solution, where one reads these sequences from right to left. These solutions are the following sequences:

7 5 3 8 6 9 3 5 7 4 3 6 8 5 4 9 7 2 6 4 2 8 1 2 1 9 1
3 4 7 8 3 9 4 5 3 6 7 4 8 5 2 9 6 2 7 5 2 8 1 6 1 9 1
3 4 7 9 3 6 4 8 3 5 7 4 6 9 2 5 8 2 7 6 2 5 1 9 1 8 1

It is certainly not easy to find them, but it is easy to check that all the constraints are satisfied: each integer appears thrice, and there are seven numbers between the first and the second occurrences of 7, and also seven numbers between the second and the third ones. This constraint is also satisfied for the eight other integers. The other problems of the family may require the placing of fewer numbers, such as the integers from 1 to 6, or more with the integers from 1 to 13. One can also change the number of times that each integer must appear: it was three in the preceding example, but one can also ask that each number appears four times.

Solving problems with the combinatorial method

Having just given some examples of CSPs, we will now examine how they can be solved. The first idea is to use a purely combinatorial method: in these problems, one must find the values of several variables so that all the constraints are satisfied. If each variable can only take a finite set of values, one can choose a variable, give it one of its possible values, then choose a second variable, give it one of its possible values, and carry on with this process until each variable has a value. Then one can check whether all the constraints are true. If this is the case, one has found a solution. If at least one constraint is false, this is not a solution, and one considers the next value for the last variable that has been chosen. If all its values have been considered, one will consider the following value of the last but one variable, and so on until one has considered all the possible values of the first variable. We have seen with crosswords that considering the following possible value of the last variable is called "to backtrack". With such a combinatorial search, one is certain to find all the possible solutions. In this way, one generates a tree, where the leaves are solutions, or contradictions when one constraint is not satisfied.

Naturally, it is possible to improve this combinatorial method. In particular, if one has given a value to all the variables of a constraint, one checks whether this constraint is false; if this happens, one immediately backtracks without waiting to instantiate all the waiting variables. In that way, one avoids wasting a lot of time; as it is better that this occurs frequently, the order chosen for instantiating the variables is essential. For instance, it is better to first consider the variables that appear in constraints with few variables, or those that appear in many constraints. Another improvement comes from the fact that one can directly find the value of a variable when other variables are known. This is the case in the crypto-addition DONALD where T is known if the value of D is known, since 2*D=T or 2*D=T+10: if the value of D is 3, the value of T is 6, and if the value of D is 7, the value of T is 4.

The combinatorial approach is convenient for a computer because it is easy to program it, and the current fast computers can consider a huge number of possibilities in a short time. They can develop trees with billions of leaves, and developing a tree with only one million leaves is almost instantaneous. This method is particularly effective when it is cleverly made, that is the order for instantiating the variables is well chosen. CAIA has the possibility to solve a problem in writing a combinatorial program; this does not give an elegant solution, but may be fast. Anyway, it is interesting to compare the results of a combinatorial program with those obtained when it uses the cleverer method that we will describe now. All in all, CAIA solves with this method any crypto-addition in less that one second; in that time, it writes the program, it compiles it, and it executes it. However, for some problems where the space search is huge, the combinatorial method may require so much time that it is unusable: even when the program is carefully written, it may require centuries of computer time.

Using a clever method for solving problems

Usually, humans do not solve problems with this method, except for the rare cases where the space search is so small that they can develop it easily. We are not very good at developing trees, even the best problem solvers make mistakes. For instance, to solve the problem of putting eight Queens on a chessboard in such a way that a Queen could not capture another Queen, Gauss believed that there were only 74 solutions, when any computer program immediately finds the 92 solutions. For that reason, we prefer to use another method, which I will call the clever method. It was developed by Jean-Louis Laurière more than 30 years ago. CAIA can also use this method, and it often finds solutions more elegant than those found by excellent human problem solvers. In this method, one tries either to find the solutions directly, without any backtrack, or to decrease the size of the search space, so that the combinatorial method becomes acceptable, even for men who cannot consider many possibilities. This does not mean that those who write a combinatorial program are not intelligent, on the contrary they have to be very clever to write an efficient combinatorial program. However, in this case, it is the programmer who is intelligent, not the program. With the clever method, the program is very intelligent because it develops a very small tree, and often no tree at all.

Let us consider some of the problems described at the beginning of this appendix, for instance the crypto-addition MONEY. When we look at the left column, we see that M is a carry, so its value is 0 or 1. As a constraint states that the value of the first letter of a line is different from 0, that means that the value of M is 1, the only remaining value. Then when we look at the second columns from the left, we see that c+S+1=O+10, c being the carry from the column at its right. So, c+S=O+9. As the value of O is not 1, a value already taken by M, the value of O+9 is 9 if O=0, or a number greater than 10 if O is greater than 1. On the other hand, as the value of the carry c is 0 or 1, and the value of S is at most 9, the value of C+S is at most 10. As c+S is equal to O+9, while the value of the first element is at most 10, and the value of the second is 9 or a number greater than 10, the only common value is 9. So, O+9=9, which gives O=0, and c+S=9. This last equality gives two possibilities: c=0 and S=9, or c=1 and S=8. I will not describe the following steps, but one can arrive directly at the only solution, without backtracking. For many problems, it is impossible to find the solutions without backtracking, but one does it in a very limited way. CAIA solves the problems in reasoning in the same way as I did in this paragraph, and it can explain its solutions as I do.

Some effective tricks for solving problems

Many other tricks are useful for solving CSPs. For the ARROWS family of problems, one can see that the value of a square where the arrow points toward the outside of the rectangle is zero, since there is no square in this direction. In the same way, when the arrow points only toward one square, its value is certainly 1 because the set of possible values contains exactly one element.

For the Rook's tour, one can see that some squares only have two neighbors, for instance in the corners of the board. As each square is connected to two other squares in a circuit, we are sure that a solution will contain the two only links between such a square and its two neighbors. In that way, we can build parts of the circuit which are obligatory; one has only to find how to connect the other squares.

Some rules can reveal a contradiction, which means that the present sub-problem has no solution: one has to backtrack immediately without wasting more time on an impossible sub-problem. For example, let us assume that there are 20 pigeons and 19 pigeon-holes which can contain only one pigeon. The problem is to assign each pigeon to a pigeon-hole, and it is evidently impossible. One does not need to try all the possible assignments, a rule states that there is a contradiction if there are more elements in the departure set than in the arrival set, and if the elements must have different values. Thus, CAIA immediately finds that there is no solution; for its part, a combinatorial program can prove this impossibility after considering all the possible assignments of the pigeons. It succeeds, because there are only 20 pigeons, but if there were 1,000,000 pigeons and 999,999 pigeon-holes, it would be almost impossible to consider all the possibilities, while CAIA still finds instantly that there is no solution. An intelligent system which uses rules for examining the formulation of a problem can therefore have a tremendous advantage over a purely combinatorial program.

Example of a rule

CAIA uses almost 200 rules for its two research domains, CSP and arithmetic. We will examine one of them, which can be applied in both domains to an equality constraint where there is one term or a sum of terms on each side. If all the terms of both sides except one are divisible by an integer N, then the last term is also divisible by N. As the first example of the application of this rule, we take again the crypto-addition DONALD. The right column gives the constraint: 2*D=T+10*c, c being a carry, its value is 0 or 1. There are three terms, 2*D, T and 10*c, two are multiple of 2, so the third term is also multiple of 2: T is even. This rule eliminates half of the possible values for T.

To show the generality of this rule, we will use it for an arithmetic problem, which has only one constraint: $4*M+3*N^2=34$. One must find the values for the variables M and N which can take any positive or negative value. CAIA begins to apply the preceding rule with N=2: 4*M and 34 are even, so $3*N^2$ is also even. As 3 is odd, another rule finds that N^2 is even, then a third rule finds that N is also even. However, if N is even, N^2 is a multiple of 4, so $3*N^2$ is also a multiple of 4. Then CAIA applies the rule again: as both terms 4*M and $3*N^2$ are multiples of 4, the third term 34 must also be a multiple of 4. Unfortunately, 34 is not a multiple of 4, so there is a contradiction: this problem has no solution.

As they can have an infinite number of possible values, it is impossible to use the combinatorial method. However, we could restrict the problem in assuming that the possible values of M and N are between -1,000,000,000 and 1,000,000,000. The combinatorial method would successively consider the 2,000,000,001 possible values for N, compute $34-3*N^2$, and test whether the result is divisible by 4, which would give the value of M. Naturally, this would always fail, but a computer would waste a lot of time doing this, while CAIA instantaneously finds that there is no solution in the more general case where the values of the variables are not restricted.

How the clever method works

Efficiently using its rules, CAIA progressively reduces the size of the search space. It makes a sequence of cycles: at each cycle it chooses a rule, and applies it. The trick is not to waste its time on useless rules, and also to try first the rules which have the greatest chance of success. Thus, an important task of CAIA is to build a method for using its rules in an effective way: it defines how to choose the rules that will be considered. To do this, it examines the conditions and the results of its rules, and it also takes into account the successes and failures which occur when it experiments on their use. A part of its work as a scientist is to learn the method that the solver agent will use. I call this method the clever method because we would consider that a human being who found such elegant solutions would be very clever.

However, it may be that, after many rules, the problem is not solved, and there is no rule that can still be executed: all the rules which could be useful have been considered, the size of the search space has been reduced, but that is not enough. In that case, it temporarily comes back to the combinatorial method: it chooses a variable, and considers all its possible values. However, after this choice, it resumes its usual method, and tries to apply rules to the new situation which has been modified by the knowledge of the value of the chosen variable. This approach is similar to the way a mathematician works when he makes a proof by cases. For instance, he will consider three cases: X<0, X=0, and X>0, then he tries to solve the three new sub-problems.

The advantages of the clever method

With the clever method, the size of the tree is significantly smaller than with the combinatorial method, even when this last method is cleverly implemented. For one of the problems of the ARROWS family, the combinatorial program written by CAIA has found all the solutions after developing a tree with 4,231,909,458 leaves, when it has developed with the clever method only 776 leaves. However, the computer times necessary for solving these problems are comparable, 154 seconds for the combinatorial program, and 67 seconds for the clever method. I consider that a solution is elegant when the size of the tree is low; when the number of leaves is equal to the number of solutions, the resolution is perfect, and it is impossible to generate a smaller tree. For the preceding examples, CAIA finds a perfect solution for the magic square, the crypto-addition MONEY, and the ARROWS problem. The tree has two leaves (one solution and one contradiction) for the crypto-additions DONALD and WASTER, and eight leaves (one solution and seven contradictions) for the Rook's tour. Langford's problem is much more difficult, since its tree had 598 leaves (3 solutions and 595 contradictions).

Depending on the method utilized for choosing the rules, the solutions may be more or less elegant, especially when one method wrongly forbids use of a rule in a situation where it would be useful. CAIA may choose between the method that I improved over several years, and the method that it has itself worked out. On more than 1,000 problems, when CAIA uses my method, the trees generated for solving these problems have 20% more leaves than the trees generated when it uses its own method. With the method that it created, CAIA found more elegant solutions than with my creation: the artificial scientist did better than the human one. Another step in the bootstrapping of CAIA has therefore been made: a task which I was doing myself, building a method for using the rules, is now performed by CAIA itself.

Often, the clever approach is not the fastest one, but humans can understand and check its solutions, because it considers fewer cases. Furthermore, it is unnecessary to examine the attempts that failed; in the same way, a mathematician will spend one year proving a theorem, and the proof is just one page long. He has made many attempts, but most of them fail; when one explains a solution, it is sufficient to give those that succeed. In contrast, with the combinatorial method, the billion of attempts that have been made must all be given if we want to understand and check the solution. This is almost impossible: if the computer has made a mistake, or if there is a bug in the program, we will never realize it. This contrasts with the resolution found by CAIA for $4*M+3*N^2=34$, where the reader can easily determine whether he agrees with the proof.

CAIA solves the Pharaoh's problem

I now return to the Pharaoh's problem given in Chapter 1: one has to find four positive integers X, Y, Z, and N such that

N is a prime number

$$X^2+Y^2=Z^2$$

$$X*Y*Z=348{,}960{,}150*N$$

CAIA begins to search for symmetries in the formulation of the problem. Finding a permutation of the variables which leads to the same formulation has been formalized as a CSP. Solving this auxiliary problem, the solver finds one solution: if one swaps X and Y, while Z and N are unchanged, the constraints are the same. To eliminate this symmetry, CAIA adds a fourth constraint: $X \geq Y$.

Then, the solver begins to work on the main problem, and I give the beginning of the explanation of its solution. Considering the third constraint, it sees that 348,960,150 is a multiple of 2, but not of 4, while N is either even or a multiple of 2 since it is a prime number. So, the second member is either a multiple of 2 or a multiple of 4. For the three variables in the product of the first member either one is multiple of 4, or two are a multiple of 2, or one is a multiple of 2, the only possibilities to give a product which is a multiple of 2 or of 4, but not of 8. If I indicate for each variable its power of 2 in its integer factorization, this gives 9 possibilities for respectively X, Y, and Z:

(2,0,0), (0,2,0), (0,0,2), (1,1,0), (1,0,1), (0,1,1), (1,0,0), (0,1,0), (0,0,1)

Seven of these possibilities are incompatible with the second constraint. For instance, if we consider (1,0,0), that means that Y and Z are odd while X is a multiple of 2^1, that is 2, but not a multiple of 4 or of a higher power of 2. Considering the second constraint, $X^2=Z^2-Y^2$, the first member is a multiple of 4, but not of 8, but the solver sees that this leads to a contradiction because the second member is at least a multiple of 8: as Y and Z are odd, it can write Y=2*M+1 and Z=2*N+1, so the second member becomes 4*(M*(M+1)+N*(N+1)). But M(M+1) is even, as is N(N+1), their sum is even, therefore the second member is at least a multiple of 8, which is forbidden. Six other possibilities are eliminated with a similar reasoning, and only (2,0,0) and (0,2,0) remain.

Let us begin a proof by case, and consider (2,0,0): X is a multiple of 4, while Y and Z are odd. The first member of the third constraint is a multiple of 4, while the second member is a multiple of 2*N, so N is a multiple of 2. As N is a prime number, that means N=2, and the third constraint becomes

X*Y*Z=697,920,300

The integer 697,920,300 can be written as: $2^2*3^5*5^2*7*11*373$. The solver finds many possibilities to restrict the possible values of X, Y, and Z, they are routine calculations. For instance, as 5 is present at the power 2, no variable can have 5 at the power 1: that would mean that two variables would be multiple of 5, and from the second constraint, the third one would also be a multiple of 5. This is impossible since the first member would be a multiple of 5^3, and the second member of 5^2 only. Finally, six possibilities remain for Z: 373, 675, 1,119, 10,071, 9,325, 27,975. The solver considers two new cases: Z≤1,119, and Z>1,119. As it knows the minimum and maximum value of each variable, using them with the constraints it easily finds that there is no solution if N>1,119, while there is one solution with the first case: X=825, Y=756, Z=1,119, and N=2.

Then the solver examines in the same way the case where X is odd and Y is a multiple of 4, and it finds no solution, because of the constraint X≥Y.

The resolution made by CAIA can be compared with the two solutions given in the site which presents this problem.

More on the subject

Sir Frederic Bartlett describes the behavior of several subjects trying to solve the crypto-addition DONALD in *Thinking: An Experimental and Social Study*.

At http://www.csplib.org/, the CSPLib has an entry for the Langford's numbers, and for many other interesting constraint satisfaction problems; CAIA has solved several of them.

The clever method for solving Constraint Satisfaction Problems was described by Jean-Louis Laurière in *A language and a program for stating and solving combinatorial problems*. A simplified description of this method was also published by Jean-Louis Laurière in *Problem Solving and Artificial Intelligence*.

I give a description of CAIA in *A Step Toward an Artificial Artificial Intelligence Scientist*. This includes several other CSPs which it has solved.

Appendix 2. How to implement some aspects of consciousness

For readers who are interested in programming, I describe the implementation of several important capacities of CAIA: its adjustable self-observation, its capacity to fully examine its present state, and to dynamically modify its behavior. Naturally, as always with computers, there are many other ways to implement them. I will only present what has really been implemented. We begin with presenting MACISTE, the infrastructure software which is the basis of CAIA. Then, we see how the programs of CAIA are written by CAIA itself, so that it can observe itself. Next, we examine how CAIA can completely observe its state when it stops. We will show that it can modify itself, not only by adding new knowledge, but by writing new programs which proceduralize these changes. Finally, we will consider the programs that have not yet been written by CAIA itself, and we will explain why I kept them, although they restrict its capacity for auto-observation.

Presenting MACISTE

I now introduce another agent, MACISTE, which I have kept hidden until now in this book so as not to confuse the reader who has no special interest in the finer points of computing. MACISTE is an infrastructure software for CAIA working under Linux. MACISTE is made up of a set of expertise fields, themselves sets of conditional actions. A conditional action has a set of conditions; when all its conditions are true, its actions must be executed. Thus, this knowledge is partially declarative: sets are completely declarative, since they do not indicate an order on their elements, but the obligation to execute a conditional action when all its conditions are true is a procedural aspect. We can see the difference from the rules used for solving a problem; when their conditions are true, they may (and not must) be executed, but one is not obliged to do so. That is why it is necessary to construct a method for using them, which adds procedural knowledge so that one knows how to use them. MACISTE has a language for expressing conditions and actions; the conditional expressions are written in a language similar to the usual mathematical language. The conditions are also expressed in a declarative form: when we have in a condition $X \in E$, MACISTE can proceduralize it in two ways. Firstly, it can consider that this defines the variable X as belonging to set E, and it will loop on all the possible values of X. Secondly, it can consider that this is a test, and it will check that the value of variable X is an element of set E. Naturally, to use the second possibility, there must be in the conditions other means for defining the value of X. Anyhow, MACISTE completely proceduralizes its expertise fields and its conditional actions before using them.

This necessity of proceduralizing the declarative aspect of MACISTE expertise fields led to the realization of an expertise, also written in its language, which translates any expertise, including itself, into an efficient C program. The C language has been chosen because it is a powerful language, reminiscent of the computers of the 1980s, and it is portable (if used with care) to many common computers; good free compilers exist for it. CAIA does not include any line of C that has not been written by CAIA itself, since it translates in C even the expertise that translates an expertise in C. MACISTE translates the 10,000 conditional rules of CAIA into 350,000 lines of C. Naturally, there was a bootstrapping problem: at the beginning, I had to write some of the procedures in C myself, but they were soon replaced by procedures translated by the first version: it has now been 20 years since the C version of CAIA has included a single line written by a human programmer. Furthermore, many other activities are managed by MACISTE: management of main memory and long term disk store, user interaction, interface to the Linux operating system, and running external programs (particularly the C compiler). All the knowledge of CAIA is not given as sets of conditional actions, it is often given in a more declarative way, such as the definition of a problem or the rules which are used to find elegant solutions. In that case, an expertise translates it into sets of conditional actions, so that it will be easily translated into C programs. It results that many conditional rules that constitute CAIA have been written by CAIA itself. As CAIA can examine its conditional actions, as well as its various kinds of completely declarative knowledge, none of its parts can escape its observation.

We have seen that a language describes the logical expressions of the conditional actions; this language is also used to express completely declarative knowledge, which is not given as conditional actions. Some expertise fields of MACISTE are particularly important because they make it possible to use in this language a MATCH function for pattern matching, and a CREATE function for creating expressions: with these functions, it is easy to analyze knowledge, and to create new knowledge. When CAIA studies itself, these functions are particularly necessary when it generates new methods for utilizing its declarative knowledge, for instance for examining the rules used for solving problems. As one has to analyze the conditions of rules, which frequently use the MATCH function, this function is often applied to expressions which contain the MATCH function itself! In the same way, the function CREATE often creates expressions which include the CREATE function. It is therefore possible to analyze knowledge that analyzes knowledge, and to create knowledge that creates knowledge; both qualities are essential for a system which observes and modifies its own behavior.

How MACISTE programs

As for normal C programs, the programs generated by MACISTE from the expertise of CAIA are compiled and link edited; this creates an executable program which is launched at the beginning of a life of CAIA. MACISTE uses a call stack: at any time, only one topmost process is running, it has been called by other processes, each having a stack frame (or activation record). Since stack frames are managed by the C compiler itself and the processor, and are not accessible to C code, MACISTE reifies its own frames on several dedicated stacks: one for local values, another for argument references, and another for return addresses. MACISTE use a call by reference mechanism. This reification (i.e. explicit representation of every call frame) is similar to continuations or reflection in many very high level language implementations (Scheme, Smalltalk, some ML, Lisp or Prolog dialects, and Java or C# reflection facilities). Therefore, CAIA knows everything of its present state, and can introspect it easily.

The call of a particular program is made by a pointer to a function without explicit argument: *f[N](); the addresses of its arguments are put into a stack by the calling process, just before the call to the function. N is often an integer, f[35] designates a particular subroutine. However, it is possible that N is an expression, so that one can dynamically indicate the subroutine that will be called at one particular place. In that case, one can execute any program, with any number of arguments since the arguments are stacked by the calling subroutine. After the completion of the execution of a process, the calling process will use the values of the arguments that have been defined, since it is a call by name. This looks complicated, but it is really very simple because all this is automatically done by MACISTE, which includes in the C programs all the instructions necessary for calling the functions, and for managing and using the stacks. When all the programs are written in such a way, it can know everything of itself when it stops, except what C instruction has just been executed when it stopped. Yet, we have seen that this can be found when it is important: it can find that it stopped in a program P, so it executes the same task again, but it writes, compiles and loads a new version of P where it adds a new instruction before every C instruction. This instruction defines the value of a new variable v: it puts v=1 before the first instruction, and v=k before the *k*th instruction. When it restarts the execution, and stops at the same point, the value of v indicates the last executed instruction.

Another interest of managing the stack of the function calls is that one knows its present size. MACISTE always tries to derecursivate its programs, so that it restricts the number of stacked calls, but a bug can lead to an abnormally long sequence of calls. Knowing the present size of this stack, the monitor agent can detect such an anomaly, and stop the task after keeping all the information necessary for analyzing this anomaly. This difficulty frequently occurred during the debugging phase of CAIA, but never during one of its lives. Anyway, even if this occurred in one of its lives, this would not prevent it from continuing, there would only be tasks that it could not complete, but it would have retained all the characteristics of these incidents.

An adjustable interpreter

After having examined how CAIA can know everything on its present state, we will show how it can interpret only the program that one wants it to examine. At the beginning of an execution, one indicates in a global variable P the name of the current target. When MACISTE writes a program, it puts at its head a few instructions which test whether this program is P. It that is true, it checks whether possible conditions on the arguments of P are satisfied: it is possible to add some conditions on the arguments of the targeted program. For instance, one can decide to interpret P which has an argument HUMAN representing a human, only if HUMAN is male, is more than 60 years old, and with a height less than 1.60 meters. The user can dynamically modify these conditions as they are interpreted by another interpreter which can only evaluate conditions. If no condition is false, the interpreter is called; its arguments are the name of the subroutine, and the index in a stack where it will find the arguments. Thus, the interpreter will begin the interpretation of P, and returns the control to the process which called P when the interpretation is completed.

If, during the interpretation, another subroutine is called, one can decide if one wants to interpret it or not. If one wants to interpret this new subroutine, the interpreter is called with the name and the arguments of this subroutine. If it does not want to interpret it, it is executed in its compiled form: it is easy to do it in executing *f[N](), N corresponding to this subroutine. However, the interpreter begins with adding the arguments in a stack, and also seeks the results in this stack when the execution of this subroutine is completed. In both cases, the interpretation of P will continue after the execution of this subroutine. At any moment during an interpretation, one can decide to stop any future interpretation of P, or to do it only if different conditions are true, or to interpret another program. If one decides to stop interpreting P, its future executions will be made with its compiled form, but the execution of the current subroutine will go to its end in the interpreted form.

In this way, CAIA has an adjustable consciousness, it can interpret what one needs to interpret, and dynamically change the program that is interpreted. In addition, the interpretation can be made with five levels of precision: at the lowest level, it gives the control to the user after each basic order, while at the highest level it does it only before executing a subroutine defined as important. At each stop, the user (which could be CAIA itself, but this possibility has not been used) can look at everything, and particularly the values of all the variables. When one has completed this analysis, one defines the next meeting point: the following instruction, or the next call to an important subroutine, or the beginning of another program, or even to stop the interpretation completely for this execution.

How CAIA observes itself

Most of the work on the information given to a computer is usually done by the compiler and the operating system, the user may have access to the information, but it is not easy, and the process depends on each system. Here, as everything is made by MACISTE, the user can make any convenient modification if it is needed, and this is completely portable from one computer to another. The C programs generated by MACISTE may look complex, but it writes them, and the user does not need to examine them. The only difficulty was to define the mechanism performed by the expertise fields of MACISTE which generate the C programs, but that was done once many years ago, and it perfectly works. In addition, if I want to give CAIA more possibilities for observing itself, I only have to modify the expertise fields that write C programs, or the expertise that interprets the expertise.

Theoretically, it is even possible to examine the programs that make these observations: one could know the values of their variables, and interpret the interpreter. However, this is rarely useful, these programs are well debugged. Furthermore, it is confusing to have the same program and the same variables used for two different goals, so I no longer use this possibility. Yet, CAIA takes into account that the highest programs in the stack are the programs that observe the other programs; it automatically ignores them, and goes down in the stack until it comes to the first process which is not used for observing.

How CAIA modifies itself

Another important point is the possibility to modify itself. We, humans, can change some of the rules which control our behavior, for instance, to decide to stop smoking. In the same way, an artificial being may change the value of some of its parameters; for instance, it will be more restrictive for keeping new problems. This is easy to do, there is no need to modify a program, one only changes the value of a variable. On the other hand, it is sometimes necessary to make more drastic modifications, which would require a human to change the structure of its brain, and for an artificial being to write new programs. This would be the case for us if we wished to be able to develop very large trees; it would be essential to add some millions of neurons for significantly enlarging the size of our working memory, which cannot keep more than ten elements for the luckiest among us. Such changes are possible for an artificial being, and they are indispensable when CAIA learns a better way for using its knowledge, or when it wants to solve a problem with the combinatorial method. It would be possible to interpret this knowledge, in the same way as we can use a sheet of paper to develop a tree but, as it has to be efficient, it is faster to use an adapted device. To do so, CAIA writes a new program, compiles it, dynamically loads it, and executes it without stopping. During one of its lives, CAIA writes and executes hundreds of thousands of instructions in programs which did not exist when it began this life.

Is it possible to remove the programs that were not written by MACISTE?

A last point linked to consciousness is to examine the programs used by CAIA, although it has not written them. This is important because it cannot examine them while they are running, they are black boxes. This is why I restricted as much as possible the utilization of such programs. In fact, it uses only two programs that it has not written: gcc, the C compiler, and the operating system Linux. As the difficulty of observing them when they are running is sometimes an impediment for CAIA, we will examine whether they could be removed.

While it would be possible to replace the C language by another one, such as Java (or preferably straight emission of JVM bytecode), C# (or bytecode for Mono), or academic languages specifically designed as intermediate representations (like LLVM or C--), these languages did not exist in the 1980s when I started implementing MACISTE. Conversely, C still exists and is widely available. Indeed, efficient machine code generation is still a very difficult task for compilers like gcc, and formal techniques, notably constraint satisfaction, mastered by CAIA could help in this tremendous task. However, this has not been achieved in MACISTE itself because it would be difficult to debug the system, and this would require very detailed declarative descriptions of machine instruction sets and current processors. The C compiler is an excellent program, and I had practically no difficulty when CAIA used it. The only difficulty occurred when it wrote combinatorial programs for solving CSPs with thousands of constraints: sometimes, it generated a subroutine with 100,000 C instructions, and gcc will not compile so large a program. I had therefore to modify the expertise that wrote the programs so that it automatically decomposes a large program into several programs restricted to about 10,000 C instructions. To conclude, theoretically, the C compiler could be removed, but I did not do it, and I do not intend to do it in the future because the game is not worth the candle.

On the other hand, I do not like the presence of the operating system, which is a huge black box playing a crucial part in the behavior of CAIA. As I have known the period when computers did not yet have an operating system, I considered the possibility of a computer working with an operating system of my own, where CAIA could really access everything. But I abandoned this idea; practically it is much too costly to realize, the operating system executes too many essential tasks, and too much time would be wasted for describing them as MACISTE expertise. As for the C compiler, it is better to continue to live with Linux, which is an excellent operating system. In using C+Linux, we have an abstract machine which can be emulated on most computers.

When CAIA is running on a computer, it can observe and modify itself in an extraordinary way. It remains to discover how to make full use of these outstanding capacities.

More on the subject

I gave a simplified description of MACISTE in *Implementation of a Reflective System.* I describe some MACISTE functions in *Speaking about and Acting upon Oneself*, while in *What Does "Itself" Mean?* I indicate how different versions of MACISTE coexist.

Bibliography

Wikipedia is an essential site for the deeper study of many domains considered in this book.

The reader who wants to explore various issues of Cognitive Science will find a lot of useful information in *The MIT Encyclopedia of the Cognitive Sciences*, edited by Robert Wilson and Frank Keil, 1999, The MIT Press.

Most of the papers presented at the 2007 AAAI Fall Symposium on AI and Consciousness are available at: http://www.consciousness.it/CAI/CAI.htm. Moreover, the presentation of this symposium by Antonio Chella and Riccardo Manzotti includes many references to papers on consciousness and AI.

Allwood, C. 1984. "Error detection processes in statistical problem solving", *Cognitive Science*, 8, 413–437.

Anceau, F. 1999. *Vers une Étude Objective de la Conscience*, Hermes.

Anderson, M. and Anderson S. 2007. "Machine Ethics: Creating an Ethical Intelligent Agent", *AI Magazine*, Winter, 15–26.

Asimov, I. 1950. *I, Robot*, Bantam.

Asimov, I. 1985. *Robots and Empire*, Panther.

Bartlett, F. 1958. *Thinking: An Experimental and Social Study*, Unwin University Press.

Bonasso, P. and Myers K. 1998. "The Home–Vacuum Event", *AI Magazine*, 19, Fall, 29–32.

Burgard, W., Cremers, A., Fox, D., Hähnel, D., Lakemeyer, G., Schulz, D., Steiner, W. and Thrun, S. 1999. "Experiences with an interactive museum tour-guide robot", *Artificial Intelligence*, 114, 3–55.

Byers, Z., Dixon, M., Smart, W. and Grimm, C. 2004. "Say cheese! experiences with a robot photographer", *AI Magazine*, 25, Fall,37–46.

Cipra, B. 1983. *Misteaks... and How to Find Them before the Teacher Does*, Birkhauser.

Chella, A. and Manzotti, R. 2007. *Artificial Consciousness*, Imprint Academic.

Clancey, W. 1983. "The epistemology of a rule-based expert system: a framework for explanation", *Artificial Intelligence*, 20, 215-251.

Damasio, A. 1995. *Descartes' Error: Emotion, Reason and the Human Brain*, Picador.

Davis, M. 1958. *Computability and Unsolvability*, McGraw-Hill.

de Groot, A. 1965. *Thought and Choice in Chess*, Mouton.

Dennett, D. 1991. *Consciousness Explained*, Little, Brown.

Ernandes, M., Angelini G. and Gori M. 2008. "A Web-Based Agent challenges Human Experts on Crosswords", *AI Magazine*, 29, Spring, 77–90.

Fabre, J.-H. 1900. *Souvenirs entomologiques*, volume VI. Available at: http://gallica.bnf.fr/ark:/12148/bpt6k99164h/f336.table.

Ganascia, J.-G. 2007. "Using non-monotonic logic to model machine ethics", *Seventh International Computer Ethics Conference*.

Gold, K., Doniec, M., Crick, C., and Scassellati, B. 2009. "Robotic vocabulary building using extension inference and implicit contrast", *Artificial Intelligence*, 173, 145-166.

Haikonen, P. 2003. *The cognitive approach to conscious machines*, Imprint Academic.

Haikonen, P. 2007. *Robot brains: Circuits and Systems for Conscious Machines*, Wiley.

Hofstadter, D. 1979. *Gödel, Escher, Bach: an Eternal Golden Braid*, Basic Books.

Hofstadter, D. 2007. *I am a Strange Loop*, Basic Books.

Holland, O. 2003. *Machine Consciousness*, Imprint Academic.

Horswill, I. 1998. "The Find-the-Remote Event", *AI Magazine*, 19, Fall, 25–28.

Jonsson, A., Morris, R. and Pedersen, L. 2007. "Autonomy in space: current capabilities and future challenges", *AI Magazine*, Winter, 27–42.

Kaplan C., and Simon H. 1990. "In search of insight", *Cognitive Psychology*, 22, 374–419.

Kuipers, B. and Stroupe, A. 2003. "The AAAI-2002 Robot Challenge", *AI Magazine*, 24, Spring, 65–71.

Ladrière J. 1957. *Les Limitations Internes des Formalismes*. Gauthier-Villars.

Lahlou, S. 2000. "Attracteurs cognitifs et travail de bureau", *Intellectica*, No. 30, 75–113.

Laurière, J.-L. 1978. "A language and a program for stating and solving combinatorial problems", *Artificial Intelligence*, 10, 29–127.

Laurière, J.-L. 1979. "Toward efficiency through generality", *Proceedings of the Sixth IJCAI*, 619–621.

Laurière, J.-L. 1986 (1989 for the English translation). *Problem Solving and Artificial Intelligence*. Prentice-Hall, New York. The French version is available at http://jll-ia-1987.googlecode.com/svn/trunk/jll-ia-1987/etude JUMP/

Levelt W. and Flores d'Arcais G. 1978. *Studies in the Perception of Language*, Wiley.

Lenat D. 1982. "The nature of heuristics", *Artificial Intelligence*, 19, 189–249.

Lenat D. 1983. "Theory formation by heuristic search", *Artificial Intelligence*, 21, 31–59,.

Lenat D. 1983. "EURISKO:A program that learns new heuristics and domain concepts", *Artificial Intelligence*, 21, 61–98.

Lévy, D. 2008. *Love and Sex with Robots*, Duckworth Overlook.

Littman M., Keim G. and Shazeer N. 2002. "A probabilistic approach to solving crossword puzzles", *Artificial Intelligence*, 134, 23–55.

Loebner's prize site: http://www.loebner.net/Prizef/loebner-prize.html.

Maxwell, B. et al. 2004. "2003 AAAI Robot Competition and Exhibition", *AI Magazine*, 25, Summer, 68–80.

McCarthy, J. 1959. "Programs with common sense", *Mechanisation of Thought Processes*, Her Majesty's Stationery Office, vol.1, 75-91.

McCarthy, J. 1995. "Making robots conscious of their mental states", *AAAI Spring Symposium on Representing Mental States and Mechanisms*, Palo Alto. A revised version is available at: http://www-formal.stanford.edu/jmc/consciousness/consciousness.html.

Michaud, F. and Gustafson, D. 2002. "The Hors d'Oeuvres Event at the AAAI-2001 Mobile Robot Competition", *AI Magazine*, 23, Spring, 31–35.

Minsky, M. 1988. *The Society of Mind,* Simon & Schuster.

Minsky, M. 2006. *The Emotion Machine: Commonsense Thinking, Artificial Intelligence, and the Future of the Human Mind*, Simon & Schuster.

Newell, A. and Simon, H. 1961. "GPS, a program that simulates human thought", reprinted in *Computers and Thought* (1963), ed. Feigenbaum and Feldman, McGraw-Hill, 279–293.

Nunn, J. 1992. *Secrets of Rook Endings*, Batsford.

Nunn, J. 1994–2002. *Secrets of Pawnless Endings*, Batsford.

Nunn, J. 1995. *Secrets of Minor-Piece Endings*, Batsford.

Olsen, M., Siegelmann-Danieli, N. and Siegelmann, H. 2008. "Robust artificial life via artificial programmed death", *Artificial Intelligence*, 172, 884–898.

Pinson, S. 1987. *Méta-modèle et heuristiques de jugement: le système CREDEX. Application à l'évaluation du risque crédit entreprise*. Thesis Pierre and Marie Curie University.

Pinson, S. 1989. "Credit risk assessment and meta-judgment", *Theory and Decision*, 27, 117–133.

Pitrat, J. 1976. "A program for learning to play chess", in Chen ed., *Pattern Recognition and Artificial Intelligence*, Academic Press, pp. 399–419.

Pitrat, J. 1990. *Métaconnaissance Futur de l'Intelligence Artificielle*, Hermes.

Pitrat J. 1995. *De la Machine à l'Intelligence*, Hermes.

Pitrat, J. 1995. *Speaking About and Acting upon Oneself.* Publication LAFORIA 95/29. Available at: http://webia.lip6.fr/~pitrat/Speaking_about_oneself.doc.

Pitrat, J. 1995. *What Does "Itself" Mean?* Publication LAFORIA 95/28. Available at: http://webia.lip6.fr/~pitrat/Itself.doc.

Pitrat, J. 1996. "Implementation of a reflective system", *Future Generation Computer Systems*, 12, 235–242.

Pitrat, J. 2008. *A Step Toward an Artificial Artificial Intelligence Scientist.* Available at: http://pagesperso-orange.fr/jacques.pitrat/.

Rosenbloom, P. 1950. *The Elements of Mathematical Logic*, Dover.

Rybski, P. et al. 2006. "The AAAI 2005 Mobile Robot Competition and Exhibition", *AI Magazine*, 27, Fall, 85–102.

Rybski, P. et al. 2007. "The AAAI 2006 Mobile Robot Competition and Exhibition", *AI Magazine*, 28, Summer, 101–114.

Sabah, G. 1997. "Consciousness: a requirement for understanding natural language", in Ó Nualláin S., Kevi P., and John Benjamins J. eds., *Two Sciences of Mind*, pp. 361–392.

Schoenfeld, A. 1985. *Mathematical Problem Solving*, Academic Press.

Shtulman, A. 2008. "Qualitative differences between naïve and scientific theories of evolution", *Cognitive Psychology*, 52, 170–194.

Simmons, R. 1998. "The Find-Life-on-Mars Event", *AI Magazine*, 19, Fall, 19–24.

Simmons, R. et al. 2003. "GRACE: An Autonomous Robot for the AAAI Robot Challenge", *AI Magazine*, 24, Summer, 51–71.

Simon, H. 1946. "The proverbs of administration", *Public Administration Review*, 6, 53–67.

Sloman, A. and Croucher, M. 1981. "Why robots will have emotions", *IJCAI 1981 Conference.* Available at http://www.andrew.cmu.edu/user/gholling/phil/Sloman.pdf.

Smart, W, et al. 2005. "The 2004 Mobile Robot Competition and Exhibition", *AI Magazine*, 26, Summer, 25–35.

Steels, L. 2003. Language Re-Entrance and the "Inner Voice", *Machine Consciousness*, ed. Holland, Imprint Academic, 173-186.

Weld, D. and Etzioni, O. 1994. "The First Law of Robotics (a call to arms)", *AAAI 94 Conference*, 1042–1047. Available at: http://www.cs.washington.edu/homes/etzioni/papers/first-law-aaai94.pdf.

Weizenbaum, J. 1966. "ELIZA--A Computer Program For the Study of Natural Language Communication Between Man and Machine", *Communications of the ACM*, 9, 35–36. Also at: http://i5.nyu.edu/~mm64/x52.9265/january1966.html.

Vermersch, P. 1999. "Introspection as practice" in the special issue "The view from within" of the *Journal of Consciousness Studies*, 6, 15–42.

Yanco, H. 1998. "The "Hors d'Oeuvres Anyone?" Event", *AI Magazine*, 19, Fall, 33–37.

Zeller, A. 2006. *Why Programs Fail,* Elsevier.

Finally two sites describe many problems:

Arithmetic Problems: http://www.diophante.fr/pages/arithm2.htm#hautdepage.

The CSPLib for Constraint Satisfaction Problems: http://www.csplib.org/.

Index

Printed and bound by CPI Group (UK) Ltd, Croydon, CR0 4YY
13/09/2026
14961015-0006